ASTRONC

A Self-Teaching Guide

Fourth Edition

Dinah L. Moché, Ph.D.

John Wiley & Sons, Inc.
New York • Chichester • Brisbane • Toronto • Singapore

Updated in 1996

Copyright © 1993 by John Wiley & Sons, Inc.
Published by John Wiley & Sons, Inc.

Library of Congress Cataloging-in-Publication Data

Moché, Dinah L., 1936–
    Astronomy: a self-teaching guide / Dinah Moché. — 4th ed.
      p.   cm.
    Includes bibliographical references and index.
    ISBN 0-471-53001-8 (paper)
    1. Astronomy.   I. Title.
  QB45.M723   1993
520—dc20   92-18618

CREDITS

Photographs are courtesy of the following organizations and individuals:

California Association for Research in Astronomy: 2.15

C.S.I.R.O.   2.17

Hale Observatories: I4; 6.4; 6.9; 6.16; 7.2; 9.28; 11.5; 11.7

Kitt Peak National Observatory: 4.7a; 5.1

Leiden Observatory: 6.10

Lowell Observatory: 9.22

NASA: I.1; 2.12; 2.19; 2.20; 4.1; 4.5; 4.12; 4.13; 5.1a; 5.9; 5.14; 6.3; 6.13; 6.14; 6.23; 7.8; 8.2; 8.14; 8.16; 9.1; 9.2; 9.6; 9.8; 9.9; 9.10; 9.16; 9.17; 9.18; 9.19; 9.20; 9.21; 9.23; 9.24; 9.25; 9.26; 9.27; 10.1; 10.5; 10.6; 10.7; 11.1; 11.4; 12.2; 12.3; 12.4; 12.7

National Optical Astronomy Observatories: 1.3; 2.6; 2.18; 4.7b; 4.8; 4.11; 5.11; 5.12; 6.1; 6.2; 6.5; 6.17; 6.20; 6.21; 6.22a

National Radio Astronomy Observatory/AUI, J. O. Burns, E. J. Schrier, and E. D. Feigelson: 6.11; 6.18; 6.19a; 6.19c

Princeton University Project Stratoscope: 4.9

Dr. Martin Schwartzchild, Princeton University: 4.10

J. William Schopf, Elso S. Barghoorn, Morton D. Masser, and Robert O. Gordon: 12.1

Tass/Sovfoto: 9.7

United States Air Force: 11.10

Tables and illustrations are adapted, redrawn, or used by permission of the following authors and publishers:

Table 1.1 and Appendix 5 from the *Observer's Handbook* (1989), with permission of The Royal Astronomical Society of Canada.

Table 2.1:   *Astronomy: Fundamentals and Frontiers*, 3rd edition, by Robert Jastrow and Malcolm H. Thompson. Copyright © 1972, 1974, 1977 by Robert Jastrow (John Wiley & Sons, New York).

Figure 2.13: Lockheed Missiles and Space Company.

Tables 3.1 (adapted); 11.1 (selected); 11.2 (selected); 11.3; Appendix 2:   *Astrophysical Quantities*, 3rd edition. Copyright © 1973 by C. W. Allen (Athlone Press, London).

Tables 6.2 (adapted) and 10.2:   *Realm of the Universe*, by George O. Abell. Copyright © 1964, 1969, 1973, 1980, by Holt, Rinehart and Winston, Inc. Copyright © 1976 by George O. Abell. Used by permission of Holt, Rinehart and Winston, Inc.

Table 10.3 and Appendix 1:   Adapted by George Lovi.

TO THE READER

Astronomy is a self-instructional book with a unique approach to teaching this fascinating subject. It is designed so that you can easily and quickly learn basic principles and contemporary topics by yourself. You can use the book alone as an introduction to astronomy and space exploration or you can use it as a supplement to conventional textbooks, computer software, or other methods of instruction. *Astronomy* covers clearly the topics that are most often presented in a college level course in introductory astronomy.

SPECIAL FEATURES

This text has many features that will make it particularly useful to you. The writing is lively and easy to understand. Many graphics help make technical ideas clear. Mathematics is not required. The index has been expanded so the text can serve as a glossary of astronomy and space exploration. Notable men and women who made historic discoveries are included.

You are constantly and actively *involved* in learning astronomy. Removable Star and Moon Maps are provided so that you can quickly identify real examples of interesting sky objects discussed in the text. Simple activities enable you to become more familiar with basic principles by testing these ideas on your own.

Six appendixes contain further information about the constellations, physical and astronomical data, measurements and astronomical symbols, nearby stars, and popular sky targets for small telescopes.

WHAT'S NEW IN THE FOURTH EDITION?

While keeping its predecessors' successful self-teaching format, the entire book has been updated and revised. The fourth edition incorporates *revolutionary* discoveries and *current* tools and research in the USA, Australia, Canada, Europe, Japan, and Russia plus the *best* suggestions from many readers who profitably used the first three editions.

Frontier insights into black holes, active galaxies and quasars, galaxy evolution, origin and structure of the universe, searches for extraterrestrial intelligence, and the newest Earth-based and space telescopes are described. Closeups of asteroid Gaspera, Neptune and its satellites, and the surface of

Venus show space scenes never before viewed by humans. Labeled drawings of the Keck Telescope, Compton Gamma Ray Observatory, Hubble Space Telescope data path, and Mir space station clarify space technology.

STUDY AIDS

A list of objectives for each chapter tells you instantly what information is contained there. The first time a new term is introduced, it appears in bold type and is defined. The material in each chapter is presented in short, numbered sections. Each section contains new information and usually asks you to answer a question or asks you to suggest an explanation, analyze, or summarize as you go along. You will always see the answer to the question right after you have answered it. If your answer agrees with the book's, you understand the material and are ready to proceed to the next section. If it does not, you should review some previous sections to make sure you understand the material before you proceed.

A self-test at the end of each chapter lets you find out fast how well you understand the material in the chapter. You may test yourself right after completing a chapter, or you might take a break and then take the self-test as a review before beginning a new chapter. Compare your answers with the book's. If your answers do not agree with the printed ones, review the appropriate sections (listed next to each answer).

USEFUL RESOURCES

Sources of excellent astronomy materials, activities, and references are included to enable you to become more involved with astronomy in your leisure time or career. Here you will also find a list of other books for stargazers of all ages by the author, Dinah L. Moché, Ph.D.

The author and publisher have tried to make this book accurate, up-to-date, enjoyable, and useful for you. It has been read by astronomers and many students who have contributed helpful suggestions during the preparation of the final manuscript. If, after completing the book, you have suggestions to improve it for future readers, please write to the author: Dinah L. Moché, Ph.D., c/o Professional & Trade Group, John Wiley & Sons, Inc., 605 Third Avenue, New York, NY 10158.

ACKNOWLEDGMENTS

I am especially grateful to my numerous students and lecture audiences and to readers of earlier editions of *Astronomy* whose questions and comments shaped the fourth edition.

Special thanks for always enthusiastically sharing the wonder and excitement of space with me to:

My home galaxy of stars Mollie and Bertram A. Levine; Elizabeth, Stephen, and Lucy Schwartz; and Rebecca, Rick, Cindy, Jessica, and Caroline Kahlenberg.

My counselor Ernest Holzberg, Esq. and friend Bonnie Brown.

The National Science Foundation Faculty Fellowship in Science awarded to me made possible advanced studies in astronomy.

Many luminaries have encouraged, influenced, and supported my work. I thank:

Steve Maran, American Astronomical Society; Maria Pallante, Authors Guild; Bob Finn, California Institute of Technology; Richard Dannay, Esq.; Pat Peterson, de Grummond Collection, University of Southern Mississippi; Carol R. Leven, Freelance Administrator; Laurence A. Marschall, Gettysburg College; Nicholas L. Johnson, Kaman Sciences Corporation; Mary Beth Murrill, W. M. Keck Observatory; Keith Mordoff, Lockheed Missiles & Space Company, Inc.; Richard Jackson, Bill Santoro, Joe Schank, Mamaroneck Post Office; Constance Moore, Althea Washington (Headquarters), Alan S. Wood, Kimberly Lievense, Sharon Miller, Mary Hardin, Ed McNevin, Jurrie van der Woude, Gil Yanow (JPL), Charles Borland, Billie A. Deason, Lisa Vazquez (JSC), Allen Kenitzer (MSFC), Ray Villard (STSI) of the National Aeronautics and Space Administration; Emma Hardesty, Karie Myers, National Optical Astronomy Observatories; Director Paul A. Vanden Bout, Patrick C. Crane (VLA), Pat Smiley, National Radio Astronomy Observatory; Roy Bishop, *Observer's Handbook*; Gloria Lubkin, *Physics Today*; Jacqueline Mitton, Royal Astronomical Society (U.K.); David Okerson, Science Applications International Corporation; George Lovi, Sky and Telescope columnist; Preston J. Campbell, TRW Federal Systems Division; John Percy, University of Toronto; Jay Pasachoff, Williams College.

Senior Vice President Stephen Kippur, and Kitt Allan, Patty Aitken, Mary Donovan, Richard Freese, Judith McCarthy, Barbara Mele, Clifford Mills, and Grant Totten at John Wiley & Sons, Inc.

Additionally, Third Edition: I. Robert, Victor and Esther Rozen; Jack Flynn, Andrew Fraknoi, Juliana Ver Steeg, Astronomical Society of the Pacific; Director Sidney Wolff, Carl A. Posey, and Jeff Stoner, Kitt Peak National Observatory; Elyse Murray, Bernard Oliver, and Charles Seeger (Ames), Donald K. Yeomans (JPL), NASA; Ronald Ekers, Arnold H. Rots, and

Don L. Swann, NRAO/VLA; Tobias Owen, SUNY/Stony Brook; Larry Esposito, University of Colorado; and Paul W. Hodge, University of Washington.

Second Edition: Lloyd Motz and Chien Shiung Wu, Columbia University; Harry L. Shipman, University of Delaware; Frank E. Bristow (JPL), Les Gaver, David W. Garrett, Curtis M. Graves, William D. Nixon (Headquarters), Peter W. Waller (Ames), and Terry White (JPL), NASA; Janet K. Wolfe, National Air and Space Museum; Richard W. West, NSF; Henry D. Berney, Thomas Como, Donald Cotten, Julius Feit, Sheldon E. Kaufman, Valdar Oinas, Robert Taylor, and Kurt R. Schmeller, Queensborough Community College of CUNY; and Arnold A. Strassenburg, SUNY/Stony Brook.

CONTENTS

LIST OF TABLES

INTRODUCTION: COSMIC VIEW

Strange is our situation here upon Earth.
Each of us comes for a short visit, not knowing why,
Yet sometimes seeming to divine a purpose.

Albert Einstein (1879–1955)

On a clear night in a place where the sky is really dark, you can see about 2000 stars with your unaided eye. You can look trillions of kilometers into space and peer thousands of years back into the distant past.

As you gaze at the stars you may wonder: What is the pattern or meaning of the starry heavens? What is my place in the vast cosmos? You are not alone in asking these questions. The beauty and mystery of space have always fascinated people.

Astronomy is the oldest science—and the newest. Exciting discoveries are being made today with the most sophisticated tools and techniques ever available. Yet dedicated amateurs can still make important contributions.

This book will teach you the basic concepts of astronomy and space exploration. You will more fully enjoy observing the stars as your knowledge and understanding grow. You will be able to read more deeply into the literature on topics that intrigue you, from ancient astronomy to the latest astrophysical theories and spaceflights.

As you teach yourself astronomy, refer to:

 The star maps and Moon map at the back of this book. These special, easy-to-read maps will help you locate and identify particularly interesting objects in the sky.

Simple activities you can do that demonstrate a basic idea.

Now, begin reading about the enormous tracts of space and time we call the universe, and stretch your mind!

Figure I.1. Earth photographed from space. Sunshine dramatically spotlights Earth's blue ocean, reddish-brown land masses, and white clouds from the Mediterranean Sea area to the Antarctica polar ice cap.

Our home is planet Earth, a rocky ball about 13,000 km (8000 miles) in diameter suspended in the vastness of space-time (Figure I.1).

Earth belongs to the solar system (Figure I.2). The solar system consists of one star—our Sun—plus nine known planets with their moons, asteroids, comets, and dust particles, all of which revolve around the Sun. The solar system is about 12 billion km (8 billion miles) across.

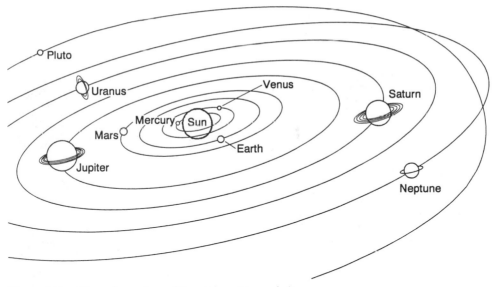

Figure I.2. The solar system. (Drawing not to scale.)

The Sun and the solar system are located in one of the great spiral arms of the Milky Way Galaxy (Figure I.3). Our immense Milky Way Galaxy includes over 200 billion stars plus interstellar gas and dust, all revolving around the center. The Milky Way Galaxy is about 100,000 light-years across. (One light-year is practically 10 trillion km, or 6 trillion miles.)

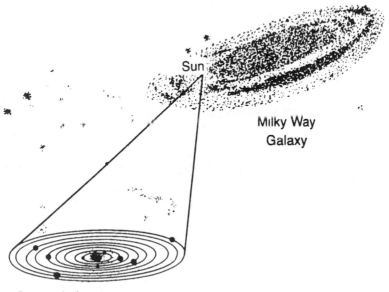

Figure I.3. The solar system in the Milky Way Galaxy.

Our Milky Way Galaxy is only one of billions of galaxies that exist to the edge of the observable universe, some 15 billion light-years away (Figure I.4).

Figure I.4. The fuzzy-looking objects in the photograph are distant galaxies. Each one includes billions of stars.

1

UNDERSTANDING THE STARRY SKY

And that inverted bowl we call the Sky
Where under crawling coop't we live and die
Lift not your hands to it for help—for It
As impotently rolls as you and I.

Rubáiyát of Omar Khayyám (1048–1131)

Objectives

☆ Locate sky objects by their right ascension and declination on the celestial sphere.
☆ Identify some bright stars and constellations visible each season.
☆ Explain why the stars appear to move along arcs in the sky during the night.
☆ Explain why some different constellations appear in the sky each season.
☆ Explain the apparent daily and annual motions of the Sun.
☆ Define the zodiac.
☆ Describe how the starry sky looks when viewed from different latitudes on Earth.
☆ Define a sidereal day and a solar day, and explain why they differ.
☆ Explain how astronomers classify objects according to their apparent brightness (magnitude).
☆ Explain why the polestar and the location of the vernal equinox change over a period of thousands of years.

1.1 STARGAZER'S VIEW

On a clear, dark night the sky looks like a gigantic dome studded with stars. We can easily see why the ancients believed that the starry sky was a huge sphere turning around Earth.

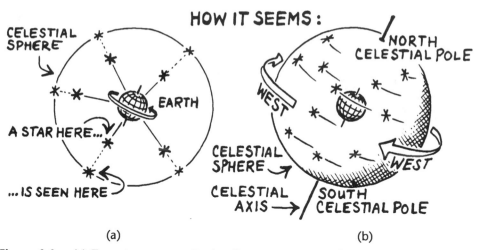

Figure 1.1. (a) To a stargazer on Earth, all stars appear equally remote. (b) We picture the stars as fixed on a celestial sphere that spins westward daily (opposite to Earth's actual rotation).

Today we know that stars are remote, blazing Suns racing through space at different distances from Earth. The Earth **rotates**, or turns, daily around its **axis** (the imaginary line running through its center between the North and South Poles).

But the picture of the sky as a huge, hollow globe of stars that turns around Earth is still useful. Astronomers call this fictitious picture of the sky the **celestial sphere**. "Celestial" comes from the Latin word for heaven.

Astronomers use the celestial sphere to locate stars and galaxies and to plot the courses of the Sun, Moon, and planets throughout the year. When you look at the stars, imagine yourself inside the celestial sphere looking out (Figure 1.1).

Why do the stars on the celestial sphere appear to move during the night when you observe them from Earth? _____

_____

Answer: Because the Earth is rotating on its axis inside the celestial sphere.

1.2 CONSTELLATIONS ⭐

It is fun to go outside and see a young blue-white star or a dying red giant star in the sky right after you read about them. You may think you will never be able to tell one star from another when you begin stargazing, but you will.

The removable star maps at the back of this book have been drawn especially for beginning stargazers observing from around 40°N latitude. (They should be useful to new stargazers throughout the midlatitudes of the northern hemisphere.)

Stars appear to belong to groups that form recognizable patterns in the sky. These star patterns are called **constellations**. Learning to identify the most prominent constellations will help you pick out individual stars.

The 88 constellations officially recognized by the International Astronomical Union are listed in Appendix 1. The more prominent, well-known ones that shine in these latitudes are shown on your star maps. Their Latin names are printed in capital letters.

Thousands of years ago people named the constellations after animals, such as Leo the Lion (Figure 1.2), or mythological characters, such as Orion the Hunter (Figure 5.2). More than 2000 years ago the ancient Greeks recognized 48 constellations.

Modern astronomers use the historical names of the constellations to refer to 88 sections of the sky rather than to the mythical figures of long ago. They refer to constellations in order to locate sky objects. For instance, saying that Mars is in Leo helps locate that planet, just as saying that Houston is in Texas helps locate that city.

Look over your star maps. Notice that the dashed line indicates the **ecliptic**, the apparent path of the Sun against the background stars. The 12 constellations located around the ecliptic are the constellations of the zodiac whose names are familiar to horoscope readers.

List the 12 constellations of the zodiac. _____

Answer: Pisces, Aries, Taurus, Gemini, Cancer, Leo, Virgo, Libra, Scorpius, Sagittarius, Capricornus, Aquarius.

(a)   (b)

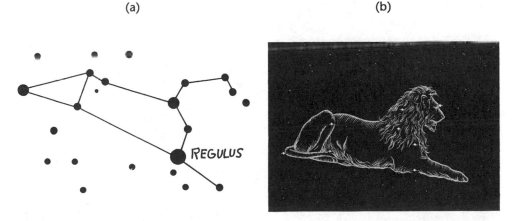

Figure 1.2. Constellation Leo is best seen in early spring when it is high in the sky. (a) Brightest star Regulus marks the lion's heart, a sickle of stars his mane, and a triangle of stars his hindquarters and tail. (b) Leo the Lion.

Figure 1.3. A time exposure taken with a camera aimed at the north celestial pole over the U.S. Kitt Peak National Observatory shows star trails that mirror Earth's actual rotation.

1.3 CIRCUMPOLAR CONSTELLATIONS

Study your star maps carefully. You will notice that several **circumpolar constellations**, near the north celestial pole (marked POLE +), appear on all four maps.

These are **north circumpolar constellations**, visible above the northern horizon all year long at around 40°N latitude (Figure 1.3). At this latitude, the south celestial pole and nearby **south circumpolar constellations** do not rise above the horizon any night of the year.

List the three circumpolar constellations closest to Polaris (the North Star) and sketch their outlines. _____

_____

_____

_____

Answer: Three circumpolar constellations that you should be able to pick out on the star maps are Cassiopeia, Cepheus, and Ursa Minor. After you know their outlines, try to find them in the sky above the northern horizon. *Note*: At latitude 40°N or higher, Ursa Major and Draco are also circumpolar.

1.4 HOW TO USE THE STAR MAPS

You can use the star maps outdoors to *identify* the constellations and stars you see in the night sky and to *locate* those you want to observe.

Choose the map that pictures the sky at the month and time you are stargazing. Turn the map so that the name of the compass direction you are facing appears across the bottom. Then, from bottom to center, your star map pictures the sky as you are viewing it from your horizon to the point directly over your head.

For example, if you are facing north about 10:00 P.M. in early April, turn the map so that the word NORTH is at the bottom. From the horizon up, you may observe Cassiopeia, Cepheus, the Little Dipper in Ursa Minor, and the Big Dipper in Ursa Major.

Name a prominent constellation that shines in the south at about 8:00 P.M. in early February. _____

Answer: Orion.

1.5 HOW TO IDENTIFY CONSTELLATIONS ★

The constellations above the southern horizon parade by during the night and change with the seasons. Turn each map so that the word SOUTH is at the bottom. Use your star maps to identify the most prominent constellations that shine each season (such as Leo in the spring and Orion in the winter).

Identify and sketch three constellations that you can see this season.

_____

_____

Answer: Your answer will depend on the season. For example, if you are reading this book in the spring, you might choose Leo, Virgo, and Boötes.

1.6 STAR NAMES ★

Long ago, more than 50 of the brightest stars were given proper names in Arabic, Greek, and Latin. The names of bright or famous stars to look for are printed on your star maps with the initial letters capitalized.

Today astronomers use alphabets and numerals to identify hundreds of thousands of stars. They refer to each of the brightest stars in a constellation by a Greek letter plus the Latin genitive (possessive) form of the constellation name. Usually the brightest star in a constellation is α, the next brightest is β, and so on. (The Greek alphabet is listed in Appendix 3.) Thus, Regulus is called α Leonis, or the brightest star of Leo. Fainter stars, not shown on your maps, are identified by numbers in star catalogs.

In a built-up metropolitan area you can see only the brightest stars. When you are far from city lights and buildings and the sky is very dark and clear, you can see about 2000 stars with your unaided eye.

Name the three bright stars that mark the points of the famous Summer Triangle. Refer to your summer skies map. _____

_____

Answer: Vega, Deneb, and Altair. Look for the Summer Triangle overhead during the summer.

1.7 BRIGHTNESS ★

Some stars in the sky look brighter than others. The **apparent magnitude** of a sky object is a measure of its observed brightness as seen from Earth. Stars may look bright because they send out a lot of light or because they are relatively close to Earth.

In the second century B.C., the Greek astronomer Hipparchus divided the visible stars into six classes, or magnitudes, by their relative brightness. He numbered the magnitudes from 1 (the brightest) through 6 (the least bright).

Modern astronomers use a more precise version of the ancient classifying system. Instead of judging brightness by the eye, they use an instrument called a **photometer** to measure brightness. Magnitudes for the brightest stars are negative—the brightest night star, Sirius, measures –1.46. Magnitudes range from –26.72 for the Sun to about +28 for the faintest objects observed in large telescopes. A difference of 1 magnitude means a brightness ratio of about 2.5.

Magnitudes are shown on your star maps and in Table 1.1. For example, we receive about 2.5 times as much light from Vega, a star of magnitude 0, as we do from Deneb, a star of magnitude 1, and about 6.3 times as much light as from Polaris, of magnitude 2. (Magnitudes are discussed further in Section 3.14.)

What do astronomers mean by "apparent magnitude"? _____

Answer: How bright a sky object looks.

1.8 LOCATION ON EARTH

The more you understand about stars and their motions, the more you will enjoy stargazing. A **celestial globe** helps you locate sky objects as a **terrestrial** (Earth) **globe** helps you locate places on Earth.

Remember how Earth maps work. We picture the Earth as a sphere and draw imaginary guidelines on it. All distances and locations are measured from two main reference lines, each marked 0°. One line, the **equator**, is the great circle halfway between the North and South Poles that divides the globe into halves. The other line, the **prime meridian**, runs from pole to pole through Greenwich, England.

Imaginary lines parallel to the equator are called **latitude** lines. Those from pole to pole are called **longitude** lines, or **meridians**. You can locate any city on Earth if you know its coordinates of latitude and longitude. Distance on the terrestrial sphere can be measured by dividing the sphere into 360 sections, called **degrees** (°). (Angular measure is defined in Appendix 3.)

Refer to the globe in Figure 1.4. Identify the equator; prime meridian; 30°N latitude line; and 30°E longitude line. (a) _____ ; (b) _____ ; (c) _____ ; (d) _____

Answer: (a) 30°N; (b) 30°E; (c) equator; (d) prime meridian.

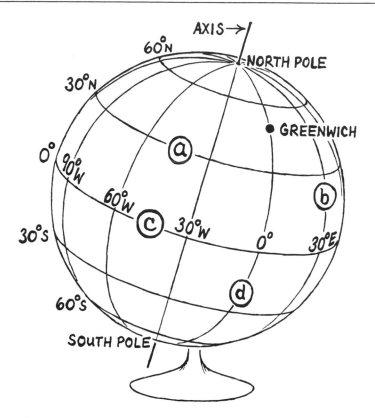

Figure 1.4. Terrestrial globe.

1.9 CELESTIAL COORDINATES

Astronomers draw imaginary horizontal and vertical lines on the celestial sphere similar to the latitude and longitude lines on Earth. They use celestial coordinates to specify directions to sky objects.

The **celestial equator** is the projection of the Earth's equator out to the sky. Angular distance above or below the celestial equator is called **declination (dec)**. Distance measured eastward along the celestial equator from the zero point, the **vernal equinox**, is called **right ascension (RA)**. Right ascension is commonly measured in hours (h), with $1^h = 15°$.

Just as any city on Earth can be located by its coordinates of longitude and latitude, any sky object can be located on the celestial sphere by its coordinates of right ascension and declination.

Give the location of the star shown in Figure 1.5. _____

Answer: 20^h RA, 30°N declination.

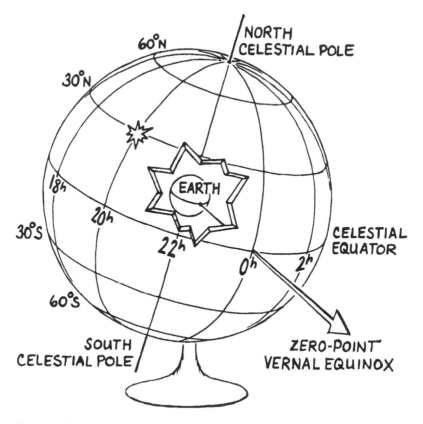

Figure 1.5. Celestial globe.

1.10 LOCATION ON THE CELESTIAL SPHERE

Every star has a location on the celestial sphere, where it appears to be when sighted from Earth. The right ascension and declination of stars for a standard **epoch**, or point of time selected as a fixed reference, change little over a period of many years. They can be read from a celestial globe, star atlas, or computer software. (See Table 1.1, for example. You'll be referring to this table when the information it contains is discussed in later chapters.)

The locations of the Sun, Moon, and planets on the celestial sphere change regularly. You can find their monthly positions in current astronomical publications and computer software (see "Useful Resources").

Explain why in any given era the stars may be found at practically the same coordinates on the celestial sphere, while the Sun, Moon, and planets change their locations regularly. _____

_____

_____

_____

Answer: The stars are too far from Earth for the unaided eye to see them move even though they are traveling many kilometers per second in various directions. The Sun, Moon, and planets are much closer to Earth. We see them move relative to the distant stars.

TABLE 1.1   The Brightest Stars

Name	Right Ascension Hours	Right Ascension Minutes	Declination Degrees	Declination Minutes	Apparent Magnitude	Spectral Class	Distance (Light Years)	Absolute Magnitude
Sun	–	–	–	–	–26.72	G2	–	4.8
Sirius	06	45.1	–16	43	–1.46	A1	8.6	1.4
Canopus	06	24.0	–52	42	–0.72	A9	74	–2.5
Alpha Centauri	14	39.6	–60	50	–0.27	G2	4.3	4.1
Arcturus	14	15.7	+19	11	–0.04	K2	34	0.2
Vega	18	36.9	+38	47	0.03	A0	25	0.6
Capella	05	16.7	+46	00	0.08	G8	41	0.4
Rigel	05	14.5	–08	12	0.12	B8	1400[b]	–8.1
Procyon	07	39.3	+05	13	0.38	F5	11.4	2.6
Achernar	01	37.7	–57	14	0.46	B3	69	–1.3
Betelgeuse	05	55.2	+07	24	0.50	M2	1400[b]	–7.2
Hadar	14	03.8	–60	22	0.61	B1	320	–4.4
Acrux	12	26.6	–63	06	0.76[a]	B2	510	–4.6[a]
Altair	19	50.8	+08	52	0.77	A7	16	2.3
Aldebaran	04	35.9	+16	31	0.85	K5	60	–0.3
Antares	16	29.4	–26	26	0.96	M2	520[c]	–5.2
Spica	13	25.2	–11	10	0.98	B1	220	–3.2
Pollux	07	45.3	+28	01	1.14	K0	40	0.7
Fomalhaut	22	57.6	–29	37	1.16	A3	22	2.0
Becrux	12	47.7	–59	41	1.25	B1	460	–4.7
Deneb	20	41.4	+45	17	1.25	A2	1500	–7.2

Source: Robert F. Garrison, *Royal Astronomical Society of Canada Observers' Handbook* (1989).

[a] Combined magnitude of double star.

[b] Distance to Orion cluster.

[c] Distance to Scorpius cluster.

Note: Parallaxes and absolute magnitudes of many stars are not well determined. For stars with a parallax smaller than 0″.05, absolute magnitudes and distances have been calculated from the spectral classification and may not be in agreement with the parallax measurement.

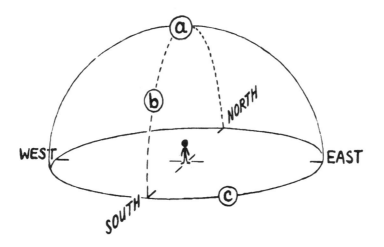

Figure 1.6. A stargazer's local reference lines.

1.11 LOCAL REFERENCE LINES

Lines of declination and right ascension are fixed in relation to the celestial sphere and move with it as it rotates around an observer. Other useful reference lines relate to the local position of each observer and stay fixed with the observer while sky objects pass by.

At your site, the **zenith** is the point on the celestial sphere directly over your head. The **celestial horizon** is the great circle on the celestial sphere 90° from your zenith. Although the celestial sphere is filled with stars, you can see only those that are above your horizon. The **celestial meridian** is the great circle passing through your zenith and the north and south points on your horizon. Only half of the celestial meridian is above the horizon.

Refer to Figure 1.6. Identify the stargazer's zenith; celestial horizon; and celestial meridian. (a) _____ ; (b) _____ ; (c) _____

Answer: (a) Zenith; (b) meridian; (c) horizon.

1.12 CELESTIAL MERIDIAN

Go outside and trace out your zenith, celestial horizon, and celestial meridian by imagining yourself, like that stargazer, at the center of the huge celestial sphere.

If possible, try this on a clear, dark, starry night. Face south. Observe the stars near your celestial meridian several times during the night. Describe what you observe. _____

_____

_____

Answer: The stars move from east to west and **transit**, or cross, your celestial meridian. This is because of the Earth's rotation from west to east. A star **culminates**, or reaches its highest altitude, when it is on the celestial meridian.

1.13 LATITUDE AND STARGAZING 💡

The stars that appear above your horizon and their paths across the sky depend on your latitude on Earth. The sky looks different from different latitudes (Figure 1.7).

If you could look at the sky from the North Pole and then from the South Pole you would see completely different stars. The Earth cuts your view of the celestial sphere in half.

You can determine how the celestial sphere is oriented with respect to your horizon and zenith at any place on Earth. In the northern hemisphere, the north celestial pole is located above your northern horizon at an altitude equal to your latitude. Polaris, the polestar, or North Star, is less than one degree away from the north celestial pole and marks the position of the pole in the sky. The declination circle that is numerically equal to your latitude passes through your zenith. In the southern hemisphere, the south celestial pole is located above your southern horizon at an altitude equal to your latitude. It is not marked by a polestar.

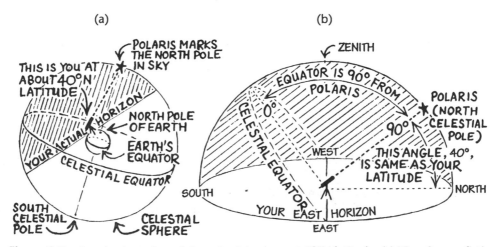

Figure 1.7.   Local orientation of the celestial sphere at 40°N latitude. (a) View from a fictitious spot on the outside. (b) Stargazer's view.

Where would you look for the North Star if you were at each of the following locations: (a) the North Pole? _____ (b) the equator? _____ (c) 40°N latitude? _____ (d) your home? _____

Answer: (a) At your zenith; (b) on your horizon; (c) 40° above your northern horizon; (d) at an altitude above your northern horizon equal to your home latitude.

1.14 APPARENT DAILY MOTION OF THE STARS

The stars appear to move in **diurnal circles,** or daily paths, around the celestial poles when you observe them from the spinning Earth.

Although the North Star, Polaris, is not a very bright star, it has long been important for navigation. Closest to the north celestial pole, it is the only star that seems to stay in the same spot in the sky. You can find Polaris by following the "pointer stars," Dubhe and Merak, in the bowl of the Big Dipper in the constellation Ursa Major. (Figure 1.8).

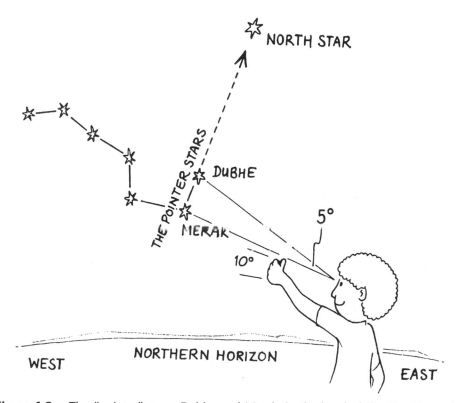

Figure 1.8.   The "pointer" stars, Dubhe and Merak, in the bowl of the Big Dipper lead you to the North Star, Polaris. The angular distance between these pointer stars is about 5° on the celestial sphere. A fist at arm's length marks about 10°. These examples will help you judge other angular distances in the sky.

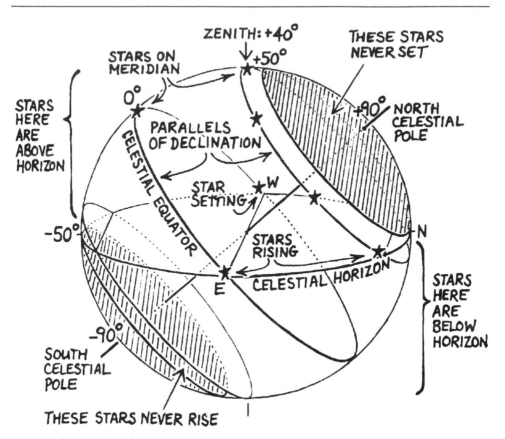

Figure 1.9. The sky from 40°N latitude. The north celestial pole is 40° above the northern horizon, and the celestial sphere rotates around it. Parallels of declination mark the stars' diurnal circles.

Since the celestial poles are at distinct altitudes in the sky at distinct latitudes, the part of a star's diurnal circle that is above the horizon is different at different latitudes on Earth. (Figure 1.9).

For example, if you stargaze at 40°N latitude, about the latitude of Denver, Colorado, U.S., you will see (Figure 1.9): (1)Stars within 40° (your latitude) of the north celestial pole (those stars between +50° and +90° declination) are always above your horizon. These stars that never set—such as the stars in the Big Dipper—are **north circumpolar stars**. (2) Stars that are within 40° (your latitude) of the south celestial pole never appear above your horizon. These stars that never rise—such as the stars in the constellation Crux, the Southern Cross—are **south circumpolar** stars. (3) The other stars, in a band around the celestial equator, rise and set. Those stars that are located at 40°N declination (equal to your latitude) pass directly across your zenith when they cross your celestial meridian.

Assume you are stargazing at 50°N latitude, about the latitude of Vancouver, Canada. Refer to Table 1.1 for the declinations of the bright stars Capella, Vega, and Canopus. Which of these stars will be above your horizon:

(a) always? _____ (b) sometimes? _____ (c) never? _____

Answer: (a) Capella (+46°00′ declination). Stars within 50° of the north celestial pole (between +40° and +90° declination) are always above the horizon. (b) Vega (+38°47′ declination). This star rises and sets. (c) Canopus (−52°42′ declination) is within 50° of the south celestial pole (between −40° and −90° declination).

1.15 UNUSUAL VIEWS

Describe how the diurnal circles of the stars would look if you were stargazing at (a) the North Pole and (b) the equator. Explain your answer. *Tip:* Remember that the celestial sphere rotates around the celestial poles. (a) _____

_____

_____

(b) _____

_____

_____

Answer: (a) All stars would seem to move along circles around the sky parallel to your horizon. The celestial sphere rotates around the north celestial pole, which is located at your zenith at the North Pole. (b) All stars would seem to rise at right angles to the horizon in the east and set at right angles to the horizon in the west. The celestial sphere rotates around the celestial poles, which are located on your horizon at the equator.

1.16 APPARENT ANNUAL MOTION OF THE STARS

The appearance of the sky changes during the night because of Earth's rotation. It also changes slowly from one night to the next.

Every night the stars appear a little farther west than they did at the same time the night before. A star rises about 4 minutes earlier each evening. Four minutes a day for 30 days adds up to about 2 hours a month. If a star is above the horizon during the daytime, the bright Sun will obscure it from view.

Thus the stars that shine in your sky at a particular time change noticeably from month to month and from season to season. In 12 months, that 4 minutes a day adds up to 24 hours. After a year, the starry sky looks the same again.

The change in the appearance of the sky with the change in seasons is due to the motion of the Earth around the Sun. The Earth **revolves**, or travels around, the Sun every year.

Picture yourself riding on Earth around the Sun, inside the celestial sphere, looking straight out. As Earth moves along in its orbit, your line of sight points toward different stars in the night sky. During a whole year you view a full circle of stars.

(a) If a star is on your zenith at 9:00 P.M. on September 1, about what time will it be on your zenith on March 1? _____ (b) Will you be able to see it? _____ Explain your answer._____

_____

Answer: (a) About 9:00 A.M. Stars rise about 2 hours earlier every month. (b) No. At that hour of the day the bright Sun obscures the distant stars from view.

1.17 THE ECLIPTIC

If the stars were visible during the day, you would see the Sun apparently move eastward among them during the year. The **ecliptic**, the apparent path of the Sun against the background stars, is drawn on sky globes and star maps for reference.

The band about 16° wide around the sky that is centered on the ecliptic is called the **zodiac**. Ancient astrologers divided the zodiac into 12 constellations, or **signs**, each taken to extend 30° of longitude (see Appendix 3). The zodiac has attracted special attention because the Moon and planets, when they appear in the sky, also follow paths near the ecliptic through these 12 constellations (Figure 1.10).

What is the zodiac? _____

_____

Answer: A belt about 16° wide around the sky, centered on the ecliptic, containing 12 constellations.

1.18 APPARENT ANNUAL MOTION OF THE SUN

The apparent easterly motion of the Sun among the stars is caused by the real revolution of Earth around the Sun. The Sun seems to move in a full circle around the celestial sphere every year.

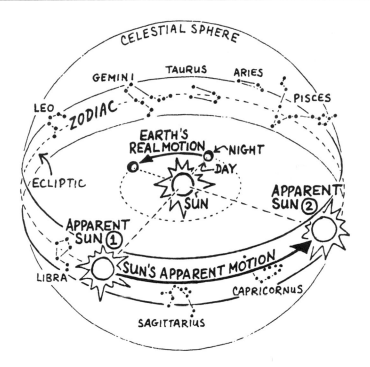

Figure 1.10. The Sun's apparent annual motion around the celestial sphere results from Earth's real motion around the Sun. As Earth orbits the Sun, different constellations of the zodiac appear in the night sky.

About how far does the Sun move on the ecliptic every day? *Tip*: Use the fact that the Sun moves 360° around the ecliptic in a year (about 365 days).

Answer: About 1°.

Solution:

$$\frac{360°}{365 \text{ days}} = 1° \text{ per day}$$

1.19 EARTH'S SEASONS

The Sun's path across the sky is highest in summer and lowest in winter. The altitude of the Sun above the horizon at noon varies during the year because Earth's axis is tilted to the plane of its orbit around the Sun (Figure 1.11).

Earth's equator remains tilted at about 23.5° to its orbital plane all year long. So as Earth travels around the Sun, the slant of the Earth–Sun line changes. Sunlight pours down to Earth from different angles during the year, causing the change of seasons as well as seasonal variations in the length of days and nights.

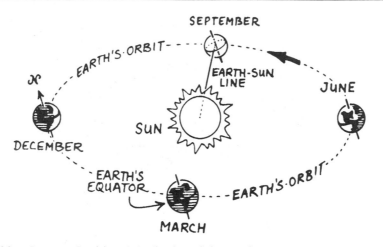

Figure 1.11. Because Earth's axis is tilted, each hemisphere gets varying amounts of sunlight during the year as our planet orbits the Sun.

Refer to Figure 1.11. Is the northern hemisphere tipped toward or away from the Sun (a) in December? _____ (b) in June? _____

Answer: (a) Away from; (b) toward.

1.20 EQUINOXES AND SOLSTICES

You can determine what the Sun's apparent position in the sky will be on any given day by checking the ecliptic on a celestial globe or a flat sky map like the one in Figure 1.12.

The **vernal equinox**, which occurs about March 21, is the Sun's position as it crosses the celestial equator going north. It is the point on the celestial sphere chosen to be the 0^h of right ascension (see Section 1.9). The **autumnal equinox**, which occurs about September 23, is the Sun's position as it crosses the celestial equator going south. At the equinoxes, day and night are equal in length.

The **summer solstice**, which occurs about June 21, and the **winter solstice**, which occurs about December 22, are the most northern and most southern positions of the Sun during the year. At these times we have the longest and shortest days, respectively, in the northern hemisphere.

Refer to Figure 1.12. Identify the vernal equinox _____; autumnal equinox _____; summer solstice _____; and winter solstice _____

Answer: vernal equinox (c); autumnal equinox (a); summer solstice (b); winter solstice (d)

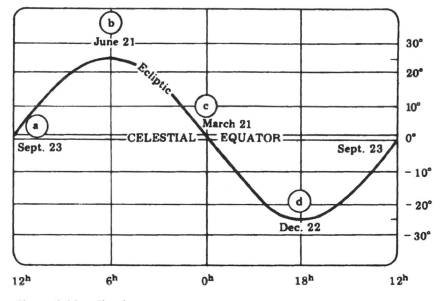

Figure 1.12. Flat sky map.

1.21 SUN'S ALTITUDE

The Sun is never directly overhead for stargazers in the midlatitudes. On a given day, the maximum altitude of the Sun in your sky depends on its declination and your latitude.

Where would you have to stand on Earth to have the Sun pass directly across your zenith at the time of the (a) vernal equinox? _____ (b) summer solstice? _____ (c) autumnal equinox? _____ (d) winter solstice? _____

Answer: (a) Equator; (b) 23.5°N latitude (Tropic of Cancer); (c) equator; (d) 23.5°S latitude (Tropic of Capricorn).

1.22 OBSERVABLE EFFECTS OF EARTH'S MOTIONS

How do the motions of Earth in space cause noticeable changes in the appearance of the sky for an observer on Earth? _____

_____

_____

_____

_____

Answer: Your summary should include the following concepts: The starry sky changes during the night because of Earth's daily rotation. The visible stars change with the seasons because of Earth's annual revolution around the Sun. The Sun's apparent daily motion across the sky is due to Earth's real rotation. The Sun's apparent annual motion is due to Earth's real revolution.

1.23 THE DAY

Earth's rotation provides a basis for keeping time using astronomical observations. The **solar day** of everyday affairs measures the time interval of Earth's rotation using the Sun for reference. The **sidereal day** measures the time interval of Earth's rotation using the stars for reference.

A sidereal day is 23 hours, 56 minutes, 4 seconds long. It is the time interval required for a star to cross your meridian two times successively, or the time for Earth to complete one whole turn in space. A solar day is 24 hours long, the length of time required for two successive meridian transits by the Sun.

A solar day is about 4 minutes longer than a sidereal day because while

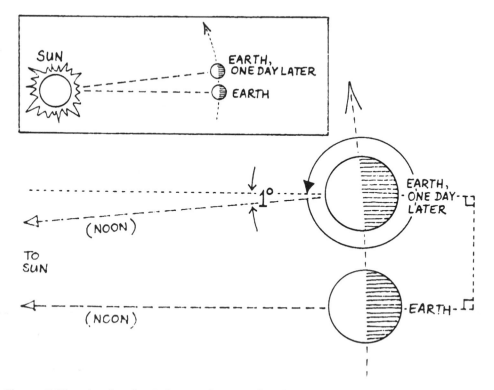

Figure 1.13. A solar day is longer than a sidereal day because during the time Earth rotates it also moves along its orbit around the Sun. In the interval from one noon to the next, Earth completes slightly more than one whole turn in space.

Earth rotates on its axis it also moves along in its orbit around the Sun. Earth must complete slightly more than one whole turn in space before the Sun reappears on your meridian. (Figure 1.13).

A clock that keeps sidereal time is useful for stargazing. In sidereal time, all stars return to their identical positions in the sky every 24 hours. So a star rises, transits the meridian, and sets at the same sidereal time all year long.

You can use celestial coordinates (see Table 1.1) to determine the sidereal time at any instant when you are stargazing. Local sidereal time is equal to the right ascension of stars on your meridian. For example, if you see brilliant Sirius transit, the sidereal time is 6 hours, 45.1 minutes.

What motion of Earth causes the 4-minute difference between a sidereal and a solar day? _____

Answer: Earth's revolution around the Sun.

1.24 PRECESSION

Your star maps will be useful to you for the rest of your life. You may be interested to know, however, that they will finally go out of date hundreds of years from now.

The direction of Earth's axis in space shifts extremely slowly around a circle once about every 26,000 years. The slow motion of Earth's axis around a cone in space is called **precession**. The precession of Earth's axis is caused

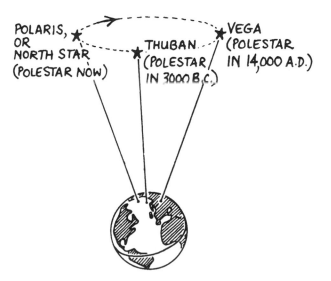

Figure 1.14. Precession. Earth's axis very slowly traces out a cone in space, so eventually the polestar changes.

mainly by the tug of the gravity of the Sun and Moon on Earth's equatorial bulge.

As Earth's axis precesses, the polestar changes. The vernal equinox, the zero point of right ascension, drifts westward around the ecliptic at a rate of about 50 seconds a year. It drifts 30°, a whole zodiac constellation, in 2150 years. Then all star charts are out of date. (Astronomers revise their precise star charts at 50-year intervals.)

In astrology today, each sign of the zodiac bears the name of the constellation for which it was originally named but with which it no longer coincides due to precession of the equinoxes.

Refer to Figure 1.14. The present polestar is Polaris, and the vernal equinox is located in the constellation Pisces. (a) What was the polestar in the year 3000 B.C.? _____ (b) What will it be in the year A.D. 14,000? _____

Answer: (a) Thuban; (b) Vega.

SELF-TEST

This self-test is designed to show you whether or not you have mastered the material in Chapter 1. Answer each question to the best of your ability. Correct answers and review instructions are given at the end of the test.

1. For each of the following references used on a terrestrial globe, list the corresponding name on the celestial sphere:

    (a) Equator. _____

    (b) North Pole. _____

    (c) South Pole. _____

    (d) Latitude. _____

    (e) Longitude. _____

    (f) Greenwich, England. _____

2. Refer to Table 1.1. Which of the five brightest stars in the sky are above the celestial equator, and which are below? _____

    _____

3. Refer to Table 1.1. Which of the five brightest stars never appear above the horizon at latitude 40° (about New York City)? _____

    _____

4. Match where you might be on Earth with the correct description of the stars:

    _____ (a) The stars seem to move along circles around the sky parallel to your horizon.

    _____ (b) The stars rise at right angles to the horizon in the east and set at right angles to the horizon in the west.

    _____ (c) Vega practically crosses your zenith.

    _____ (d) Alpha Centauri is always above your horizon.

    _____ (e) Polaris appears about 30° above your horizon.

    (1) Antarctica (below 61°S).
    (2) Equator.
    (3) Jacksonville, Florida, U.S. (30°22′N).
    (4) North Pole.
    (5) Sacramento, California, U.S. (38°35′N).

5. Why do the stars appear to move along arcs in the sky during the night?

    _____

6. Why do some different constellations appear in the sky each season?

   _____

7. What is the zodiac? _____

   _____

8. Where on Earth would you have to be to have the Sun pass directly across your zenith at the time of the (a) vernal equinox? _____
   (b) summer solstice? _____ (c) winter solstice? _____

9. If a star rises at 8 P.M. tonight, at approximately what time will it rise a month from now? _____

10. Why is a solar day about 4 minutes longer than a sidereal day? _____

    _____

11. Arrange the following stars in order of decreasing brightness: Antares (magnitude 1); Canopus (magnitude –1); Polaris (magnitude 2); Vega (magnitude 0)._____

12. Why will the polestar and the location of the vernal equinox on the celestial sphere be different thousands of years from now, causing your star maps finally to go out of date? _____

    _____

ANSWERS

Compare your answers to the questions on the self-test with the answers given below. If all of your answers are correct, you are ready to go on to the next chapter. If you missed any questions, review the sections indicated in parentheses following the answer. If you missed several questions, you should probably reread the entire chapter carefully.

1. (a) Celestial equator.   (d) Declination.

   (b) North celestial pole.   (e) Right ascension.

   (c) South celestial pole.   (f) Vernal equinox.

   (Sections 1.1, 1.8, 1.9)

2. Above: Arcturus, Vega.   Below: Sirius, Canopus, Alpha Centauri.
   (Sections 1.9, 1.10)

3. Canopus, Alpha Centauri.   (Sections 1.10, 1.13, 1.14)

4. (a) 4;   (b) 2;   (c) 5;   (d) 1;   (e) 3.   (Sections 1.10, 1.13 through 1.15)

5. Because of Earth's rotation.   (Sections 1.1, 1.12, 1.14)

6. Because of Earth's revolution around the Sun.   (Section 1.16)

7. A belt about 16° wide around the sky centered on the ecliptic, containing 12 constellations.   (Section 1.17)

8. (a) Equator;   (b) 23.5°N (Tropic of Cancer);   (c) 23.5°S (Tropic of Capricorn).   (Sections 1.19 through 1.21)

9. 6 P.M.   (Section 1.16)

10. Because, while Earth rotates on its axis, it also moves along in its orbit around the Sun. Earth must complete slightly more than one whole turn in space before the Sun reappears on your meridian.   (Section 1.23)

11. Canopus, Vega, Antares, Polaris.   (Section 1.7)

12. Because of the precession of Earth's axis.   (Section 1.24)

2

LIGHT AND TELESCOPES

Curiosity is one of the permanent and certain characteristics of a vigorous mind.

The Rambler, Samuel Johnson (1709–1784)

Objectives

☆ Describe the wave nature of light, including how it is produced and how it travels.

☆ Name the major regions of the electromagnetic spectrum from the shortest wavelength to the longest.

☆ State the relationship between wavelength and frequency.

☆ State the relationship between the color of a star and its temperature.

☆ List the three windows (spectral regions) in Earth's atmosphere in order of their importance to observational astronomy.

☆ Explain how refracting and reflecting telescopes work.

☆ Define light-gathering power, resolving power, and magnification with respect to a telescope.

☆ State the two most important factors in telescope performance.

☆ State the purpose of a spectrograph.

☆ Explain how radio telescopes work, and list some interesting radio sources.

☆ Explain why infrared telescopes are located in very high, dry sites, and list some objects they observe.

☆ Explain why ultraviolet, X-ray, and gamma ray telescopes must operate above Earth's atmosphere, and list some objects they study.

2.1 WHAT IS LIGHT?

Most of our information about the universe has been obtained through the analysis of starlight. To explain how starlight travels across trillions of kilome-

SOURCE   RECEIVER

Figure 2.1. Visualizing a light wave.

ters of empty space to waiting telescopes, astronomers picture light as a form of wave motion.

A **wave** is a rising and falling disturbance that transports energy from a source to a receiver without the actual transfer of material. Wave motion is clearly observable in the ocean. During storms, crashing ocean waves vividly reveal the energy they carry.

A **light wave** is an electromagnetic disturbance consisting of rapidly varying electric and magnetic effects. Light waves transport energy from accelerating electric charges in stars (the source) to electric charges in the retina of your eye (the receiver) (Figure 2.1). You become aware of that energy when you see starlight.

What is a wave? _____

_____

Answer: A wave is a rising and falling disturbance that transports energy from a source to a receiver without the actual transfer of material.

2.2 WAVELENGTH

Light waves are distinguished by their lengths. The distance from any point on a wave to the next identical point, such as from crest to crest, is called the **wavelength** (Figure 2.2).

The human eye responds to waves that have extremely short wavelengths.

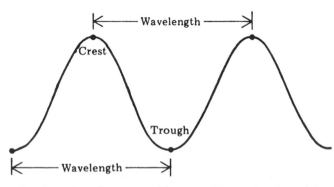

Figure 2.2. Wavelength measured from crest to crest or trough to trough.

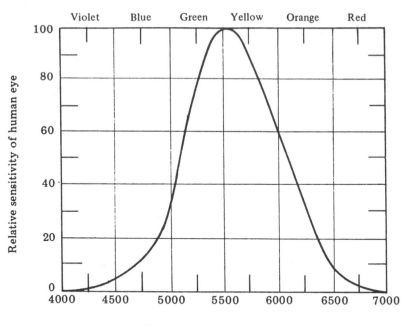

Wavelength (angstroms)

Figure 2.3. Relative sensitivity of the human eye to different colors and wavelengths of visible light.

These waves that produce vision are called **visible light** (Figure 2.3). Physicists measure these waves in **nanometers (nm)**. Astronomers commonly use the **angstrom unit (Å)**, after Swedish physicist Anders J. Ångstrom (1814–1874), who first measured wavelengths of sunlight in nanometers. One nanometer is 10^{-9} m, and one angstrom is 0.10 nm. The diameter of a human hair is about 500,000 Å!

Visible light has wavelengths of 4000 Å to 7000 Å. The varying wavelengths of visible light are perceived as different **colors**. The arrangement of the colors according to wavelength is called the **visible spectrum**.

Refer to Figure 2.3. Which color light has (a) the shortest wavelength? _____ (b) the longest wavelength? _____ (c) To which wavelength (color) is the eye most sensitive? _____

Answer: (a) Violet; (b) red; (c) 5550 Å (yellow–green).

2.3 THE ELECTROMAGNETIC SPECTRUM

Visible light is only one small part of all the electromagnetic radiation in space. Energy is also transmitted in the form of gamma rays, X-rays, ultraviolet radiation, infrared radiation, and radio waves.

Because we make such different uses of them, these forms of radiation seem very different from one another. Doctors use gamma rays in cancer treatment and X-rays for medical diagnosis. Ultraviolet rays give you a suntan, and infrared rays warm you up. Radio waves are used for communication.

Electromagnetic radiation Name of region	Wavelength (cm)	Frequency (cycles per second)
	Short	High frequency
Gamma rays		
	10^{-9}	3×10^{19}
X rays		
	10^{-6}	3×10^{16}
Ultraviolet	3×10^{-5}	10^{15}
Visible		
	10^{-4}	3×10^{14}
Infrared		
	10^{-1}	3×10^{11}
Microwaves	1	3×10^{10}
Spacecraft		
	10^{2}	3×10^{8}
Television and FM	10^{3}	3×10^{7}
Shortwave	10^{4}	3×10^{6}
	10^{5}	3×10^{5}
AM radio waves	1 km	300 kHz
	Long	Low frequency

Violet
Blue
Green
Yellow
Orange
Red

Radio waves

Figure 2.4. The electromagnetic spectrum includes all electromagnetic radiation from shortest, highest-frequency gamma rays to longest, lowest-frequency radio waves.

All of these forms of radiation are really the same basic kind of energy as visible light. They have different properties because they have different wavelengths. The shortest waves have the most energy, whereas the longest waves are the least energetic. The whole family of electromagnetic waves, arranged according to wavelength, is called the **electromagnetic spectrum**.

Electromagnetic waves of all wavelengths are important to astronomers because all such waves bring clues about their sources.

Refer to Figure 2.4. List six forms of electromagnetic radiation from the shortest waves (highest energy) to the longest waves (lowest energy). _____

_____

Answer: Gamma rays, X-rays, ultraviolet radiation, visible light, infrared radiation, radio waves.

2.4 RANGE OF WAVELENGTHS

What is the range of wavelengths included in the whole electromagnetic spectrum? _____

_____

_____

Answer: Wavelengths vary from less than a trillionth of a meter, 10^{-12} m, for the shortest gamma rays to longer than a kilometer, 10^3 m (a mile), for the longest radio waves.

2.5 SPEED OF LIGHT

All kinds of electromagnetic waves move through empty space at the same speed—that is, at the speed of light. The speed of light in empty space, usually symbolized by the letter c, is practically 300,000 km/second (186,000 miles per second).

The speed of light in empty space has been called the "speed limit of the universe," because no known object can be accelerated to move faster. It is one of the most important and precisely measured numbers in astronomy (Appendix 2).

A **light-year (ly)** is the distance light travels through empty space in one year.

How many kilometers (miles) does 1 light-year represent? *Tips*: (1) distance = speed × time. (2) A year is equal to 3.156×10^7 seconds. _____

Answer: Practically 9.5 trillion km (6 trillion miles).

Solution: Multiply 300,000 km/second $\times 3.156 \times 10^7$ second/year
(186,000 miles/second $\times 3.156 \times 10^7$ seconds/year)

2.6 WAVE FREQUENCY

Wave motion can be described in terms of frequency as well as wavelength. The **frequency** of a wave motion is the number of waves that pass by a fixed point in a given time, measured in **cycles per second (cps)**.

The human eye responds to different-color light waves that have very high frequencies. Visible light waves vary in frequency from 4.3×10^{14} cps for red to 7.5×10^{14} cps for violet, with the other colors in between.

For radio waves, one cycle per second is commonly called a **hertz (Hz)**, after the German physicist Heinrich Hertz (1857–1894), who first produced radio waves in a laboratory. An AM radio receives radio waves with frequencies of 550 to 1650 **KHz (kilohertz)**; 1 KHz is 1000 cycles per second. The FM band ranges from 88 to 108 **MHz (megahertz)**; 1 MHz is a million cycles per second.

Refer to the electromagnetic spectrum shown in Figure 2.4. Which waves have (a) a higher frequency than the visible light waves? _____
_____(b) a lower frequency than the visible light waves? _____

Answer: (a) Higher frequency: gamma rays, X-rays, ultraviolet radiation. (b) Lower frequency: infrared radiation, radio waves.

2.7 WAVELENGTH AND FREQUENCY

Can you deduce a general relationship between wavelength and frequency for these electromagnetic waves? _____

Answer: The wavelength is inversely proportional to the frequency. The shorter waves have a relatively higher frequency, and the longer waves have a relatively lower frequency.

2.8 WAVE PROPAGATION

The relationship you have just found is an example of a formula that holds true for all kinds of wave motion:

Speed of wave = Frequency × Wavelength

You can use this formula to calculate the frequency of any kind of electromagnetic wave in empty space if you know its wavelength (or the wavelength if you know the frequency). Explain why. *Tip*: Review Section 2.5. _____

_____

_____

Answer: All electromagnetic waves have the same speed in empty space—that is, the speed of light, or about 300,000 km/second (186,000 miles per second).

2.9 WAVE EQUATION

Be sure you understand the relationship between speed (c), frequency (f), and wavelength (λ) for electromagnetic waves. The formula is:

$$c = f\lambda$$

Calculate the wavelength of a radio wave whose frequency is 100 KHz (100,000 cycles per second). _____

Answer: 3 km (1.86 miles).

Solution:

Speed = Frequency × Wavelength

Thus,

$$\text{Wavelength} = \frac{\text{Speed}}{\text{Frequency}} = \frac{300{,}000 \text{ km/second}}{100{,}000 \text{ cycles/second}}$$

$$= \frac{186{,}000 \text{ miles/second}}{100{,}000 \text{ cycles/second}}$$

2.10 RADIATION LAWS ✪

Stars, like other hot bodies, radiate electromagnetic energy of all different wavelengths. The hotter the star, the more radiant energy it emits. The temperature of a star determines which wavelength is brightest.

TABLE 2.1   Four Hot and Cool Stars

Season	Star	Constellation	Color	Surface Temperature (K)
Summer	Vega	Lyra	Blue-white	10,000
Summer	Antares	Scorpius	Red	3,000
Winter	Sirius	Canis Major	Blue-white	10,000
Winter	Betelgeuse	Orion	Red	3,400

Stars radiate energy practically as a **blackbody**, or theoretical perfect radiator. **Wien's law of radiation** states that the wavelength, λ_{max} at which a blackbody emits the greatest amount of radiation is inversely proportional to its temperature (T). The formula is

$$\lambda_{max} = \frac{0.3}{T}$$

where λ_{max} is in centimeters and T is in kelvin (K). Thus the hotter a star, the shorter the wavelength at which it emits its maximum radiation.

Some stars are thousands of degrees hotter than others. You can judge how hot a star is by its color (wavelength). The hottest stars look blue-white (short wavelength), and the coolest stars look red (long wavelength). Extremely hot stars (very short wavelengths) and extremely cool stars (very long wavelengths) are not visible.

Look in the sky for the examples cited in Table 2.1.

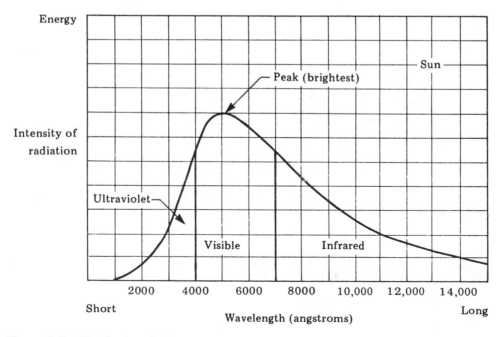

Figure 2.5.   The Sun's radiation curve.

The **Stefan-Boltzmann radiation law** states that the total energy (E), emitted by a blackbody is proportional to the fourth power of its absolute temperature (T). Thus a star that is twice as hot as our Sun radiates 2^4, or 16, times more energy than the Sun.

A **radiation curve** shows how much energy a body radiates at different wavelengths, which wavelengths it radiates most intensely, and the total amount of energy it radiates at all wavelengths (indicated by the area under the curve).

Examine Figure 2.5. (a) The Sun radiates most intensely in the _____ wavelengths. (b) The total amount of energy that the Sun radiates as visible light is (more, less) _____ than the amount radiated outside the visible region.

Answer: (a) Visible; (b) less.

2.11 ASTRONOMICAL OBSERVATIONS

Today astronomers have tools to observe and analyze all forms of electromagnetic radiation from space. The main function of a telescope—whatever type of radiation is being detected—is to gather sufficient radiation for analysis.

Figure 2.6. Kitt Peak, a 2100-m-(6900-foot-)-high site about 30 km (50 miles) outside Tucson, Arizona, U.S., has telescopes for six observatories. Included are Kitt Peak National Observatory and National Solar Observatory of the National Optical Astronomy Observatories.

Earth's atmosphere stops most radiation from space and permits only certain wavelengths to shine through to telescopes on the ground. Ground-based astronomers can look out at the universe through three **windows,** or spectral ranges, within which our atmosphere is largely transparent to radiation. These are the optical (visible light), radio, and infrared windows.

An **astronomical observatory** is a place equipped for the observation of sky objects. For ground-based observations at visible wavelengths, astronomers choose sites with frequently clear skies, on mountaintops far from city lights and pollution (Figure 2.6).

What would you suggest to astronomers who want to observe the universe in the gamma ray, X-ray, and ultraviolet ranges? _____

_____

_____

Answer: Locate their instruments beyond Earth's atmosphere. Space age technology makes space-based observations in these wavelength bands possible from rockets, spacecraft, or even Moon-based observing stations.

2.12 OPTICAL TELESCOPES

An **optical telescope** forms images of faint and distant stars. It can collect much more light from space than the human eye can. Optical telescopes are built in two basic designs—**refractors** and **reflectors**.

The heart of a telescope is its **objective,** a **main lens** (in refractors) or a **mirror** (in reflectors). Its function is to gather light from a sky object and focus this light to form an image. The ability of a telescope to collect light is called its **light-gathering power**.

Light-gathering power is proportional to the area of the collecting surface, or to the square of the **aperture** (clear diameter) of the main lens or mirror. The **size** of a telescope, such as 150-mm or 5-m (6-inch or 200-inch), refers to the diameter of its aperture.

You can look at the image directly through an eyepiece, which is essentially a magnifying glass. Or you can photograph the image or record and process it electronically. Your eye lens size is about 5 mm (0.2 inch). A 150-mm (6-inch) telescope has an aperture over 30 times bigger than your eye lens. Its light-gathering power is 30^2, or 900 times greater than that of your eye. So a star appears over 900 times brighter with a 150-mm (6-inch) telescope than it does to your unaided eye.

Astronomers build giant telescopes to detect ever fainter and more distant objects.

All stars appear brighter with telescopes than they do to the eye alone. The extra starlight gathered by the telescope is concentrated into a single point. Using time exposure, a giant 10-m (400-inch) telescope can image very faint stars down to about magnitude 28, which is the same apparent brightness as a candle viewed from the Moon!

How much brighter would a star appear with the 10-m (400-inch) telescope than to your unaided eye? Explain._____

_____

_____

Answer: Over 4 million times brighter. The 10-m (400-inch) telescope is over 2000 times bigger than your eye lens, so it gathers over 2000^2, or 4 million, times more light.

2.13 BINOCULARS

Binoculars are a practical first instrument for stargazing because they are easy to use and portable. A pair labeled 7 × 50 has an aperture of 50 mm. The 7 × specifies the magnification.

Why do binoculars and telescopes reveal many more sky objects than you can see with your unaided eye?_____

_____

Answer: They can collect much more light than your eye can. (Light-gathering power is proportional to the square of the aperture.)

2.14 REFRACTING TELESCOPES

A refracting telescope has a main, **objective lens** permanently mounted at the front end of a tube. Starlight enters this lens and is **refracted**, or bent, so that it forms an image near the back of the tube.

The distance from this lens to the image is its **focal length**. You may look at the image through a removable magnifying lens called the **ocular**, or **eyepiece**. The tube keeps out scattered light, dust, and moisture.

Italian astronomer Galileo Galilei (1564–1642) first pointed a refracting telescope skyward in 1609. The largest instrument he made was smaller than 50 mm (2 inches).

Today refracting telescopes range in size from a beginner's 60-mm (2.4-inch) version to the world's largest, the 1-m (40-inch) telescope at the Yerkes Observatory in Williams Bay, Wisconsin, U.S., which was completed in 1897.

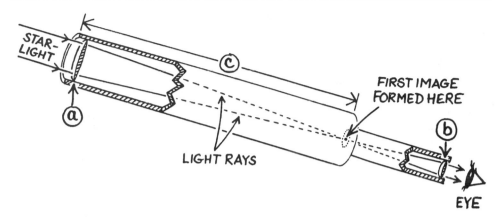

Figure 2.7. A refracting telescope with a long focal length objective lens and a short focal length eyepiece.

Refer to Figure 2.7. Identify the refracting telescope's (a) objective lens; (b) the eyepiece; and (c) the focal length of the objective lens. State the purpose of (a) and (b). (a) _____

(b) _____

(c) _____

Answer: (a) Objective lens: to gather light and form an image. (b) Eyepiece: to magnify the image formed by the objective. (c) Focal length of objective lens.

2.15 REFLECTING TELESCOPES

A **reflecting telescope** has a highly polished curved-glass mirror, the **primary mirror**, mounted at the bottom of an open tube. When starlight shines on this mirror, it is reflected back up the tube to form an image at the **prime focus**.

You can place photographic film or electronic devices at the prime focus to record the image, or you can use additional mirrors to reflect the light to another spot for viewing. The **Newtonian telescope** uses a small, flat mirror to reflect the light through the side of the tube to an eyepiece (Figure 2.8).

The **Cassegrain telescope** uses a small convex mirror, a **secondary mirror,** to reflect the light back through a hole cut in the primary mirror at the bottom end of the tube (Figure 2.9). It is more compact than a refractor or Newtonian reflector of the same aperture. The **Schmidt-Cassegrain** telescope combines an extremely short-focus spherical primary mirror at the back end of a sealed tube with a thin lens at the front.

Reflecting telescopes range in size from a beginner's 75-mm (3-inch) Newtonian reflector to the world's largest, the 10-m (400-inch) Keck Telescope atop the inactive volcano Mauna Kea in Hawaii, U.S. (Figure 2.15).

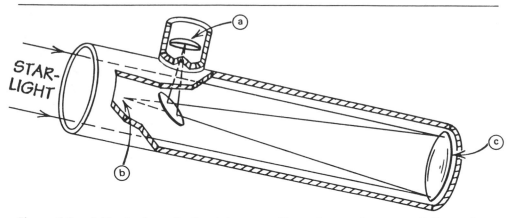

Figure 2.8. A Newtonian reflecting telescope with a primary mirror, a small diagonal secondary mirror, and an eyepiece.

Refer to Figures 2.8 and 2.9. Identify the reflecting telescope's primary mirror; eyepiece; and prime focus. (a) _____ ; (b) _____ ; (c) _____

Answer: (a) Eyepiece; (b) prime focus; (c) primary mirror.

2.16 REFLECTORS VERSUS REFRACTORS

What is the essential difference between a reflecting telescope and a refracting telescope? Explain. _____

_____

Answer: The main optical part (objective). A reflecting telescope uses a mirror, whereas a refracting telescope uses a lens to collect and focus starlight.

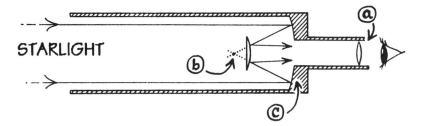

Figure 2.9. A Cassegrain reflecting telescope with a concave primary mirror, a small convex secondary mirror, and an eyepiece.

2.17 f NUMBER

Telescopes are often described by both their aperture size and f number. The **f number** is the ratio of the focal length of the main lens or mirror to the aperture. These specifications are important because the brightness, size, and clarity of the image produced by a telescope depend on the aperture and focal length of its main lens or mirror.

For example, a "150-mm (6-inch), f/8 reflector" means the primary mirror is 150 mm (6 inches) in diameter and has a focal length of 1200 mm (8 × 150), or 48 inches (8 × 6).

What is the focal length of the 5-m (200-inch), f/3.3 mirror on Mount Palomar in California, U.S.? _____

Answer: 16.5 m (660 inches, or 55 feet).

2.18 IMAGES

All stars except our Sun are so far away that they appear as dots of light in a telescope. The Moon and planets appear as small disks. **Image size** is proportional to the focal length of the telescope's main lens or mirror.

For example, a mirror with a focal length of 2.5 m (100 inches) produces an image of the Moon that measures about 2.5 cm (1 inch) across. You know that the 5-m (200-inch), f/3.3 mirror has a focal length of 16.5 m (660 inches), which is over six times as long. Hence, it produces an image of the Moon that is about six times as big, or 15 cm (6 inches) across.

Lenses and mirrors form real images that are upside down. (A real image is formed by the actual convergence of light rays.) Since inverted images do not matter in astronomical work, and righting them would require additional light-absorbing optics, nothing is done to turn images upright in telescopes.

What determines the size of the image formed by a telescope? _____

Answer: The focal length of the main lens or mirror.

2.19 RESOLVING POWER

Even if a telescope were of perfect optical quality, it would not produce perfectly focused images because of the nature of light itself. A telescope's **resolv-**

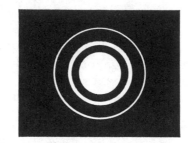

Figure 2.10. Diffraction pattern (image of a star).

ing power is its ability to produce sharp, detailed images under ideal observing conditions.

Resolving power depends directly on the size of the aperture and inversely on the wavelength of the incoming light. For the same light, a 150-mm (6-inch) telescope has twice the resolving power of a 75-mm (3-inch) telescope.

Starlight travels in straight lines through empty space, but when waves of starlight pass close to the edge of a lens or mirror, they spread out, in an effect called **diffraction**, and come to a focus at different spots. Because of diffraction, the image of a star formed by a lens or mirror appears under magnification as a tiny, blurred disk surrounded by faint rings, called a **diffraction pattern**, instead of as a single point of light (Figure 2.10). Diffraction limits resolving power.

If two stars are close together, their diffraction patterns may overlap so that they look like a single star. Features such as Moon craters and planet markings are also blurred by diffraction.

Resolving power determines the smallest angle between two stars for which separate, recognizable images are produced. The smallest resolvable angle for the human eye is about one minute of arc (1′) which is the size of an aspirin tablet seen at a distance of 35 m (110 feet).

Explain why what may look like a single star to the eye may resolve into two close neighbor stars in a telescope. _____

_____

Answer: Resolving power is proportional to aperture, and a telescope's aperture is much larger than the human eye's.

2.20 MAGNIFICATION

A telescope's **magnifying power** is the ratio of the apparent size of an object seen through the telescope to its size when seen by the eye alone. Telescopes

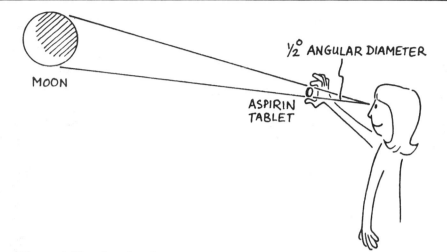

Figure 2.11. Angular diameter.

magnify the angular diameter of objects. Thus the image appears to be closer than the object.

For example, to your eye the angular diameter of the full Moon is ½°, the same as an aspirin tablet held at arm's length (Figure 2.11). If the apparent size of the Moon increases 20 times, so that it looks 10° in diameter when you view it through your telescope, then the magnifying power is 20, written 20✕.

The value of the magnifying power of a telescope depends on the eyepiece you use. You can figure:

$$\text{Magnifying power} = \frac{\text{Focal length of telescope}}{\text{Focal length of eyepiece}}$$

A telescope usually comes with several eyepieces of different focal lengths so you can vary its magnifying power for viewing different objects.

(a) What is the magnifying power of a 150-mm (6-inch), f/8 telescope when an eyepiece of 12.5-mm (½-inch) focal length is used? _____ (b) How could you increase the magnifying power of this telescope? _____

Answer: (a) 96✕.

Solution:

$$\text{Magnifying power} = \frac{\text{Focal length of telescope}}{\text{Focal length of eyepiece}}$$

$$= \frac{1200 \text{ mm}}{12.5 \text{ mm}} = \frac{48 \text{ inches}}{\text{½ inch}}$$

(b) Use an eyepiece of a shorter focal length.

(a)   (b)

Figure 2.12. Effect of atmospheric blurring on resolution. Globular star cluster M-14, located 70,000 light-years away, imaged with (a) a ground-based, 4-m telescope, Cerro Tololo Inter-American Observatory, Chile; and (b) Hubble Space Telescope. Image resolution is (a) 1.5 seconds of arc; (b) 0.08 seconds of arc.

2.21 MAXIMUM USEFUL MAGNIFICATION

It is a mistake to exaggerate the importance of magnifying power when you buy a telescope. You cannot increase the useful magnifying power indefinitely by changing eyepieces.

Starlight must pass through Earth's atmosphere to reach waiting telescopes on the ground. Disturbances in the air cause blurry images. **Seeing** refers to atmospheric conditions that affect the sharpness of a telescope's image. If the air is quiet, then the seeing is good, and stars shine with a steady light. If the air is turbulent, then the seeing is bad, and stars twinkle madly.

The **practical limit of useful magnification** for any telescope is about two times its aperture in millimeters (50 times its aperture in inches). Higher power will just magnify any blurring in the image due to diffraction or bad seeing. It cannot reveal any finer details.

A telescope in space escapes interference from Earth's atmosphere, so it can see farther and image sharper than a telescope on the ground (Figure 2.12). Astronomers can operate Earth-orbiting observatories by remote control from the ground. Astronauts can maintain a space telescope in orbit and return it to Earth for a major overhaul.

The biggest observatory in orbit around Earth is the U.S./European Hubble Space Telescope (HST), which was deployed in 1990. It has a 2.4-m (94-inch) mirror and five instruments for visible and ultraviolet observations (Figure 2.13).

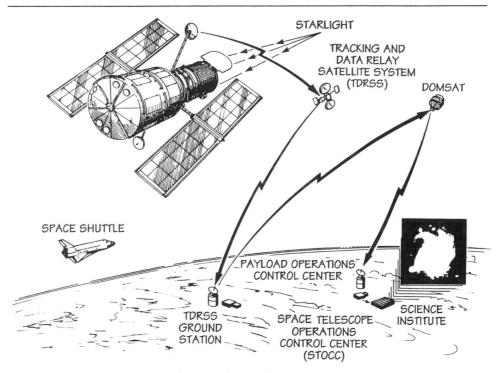

Figure 2.13. Hubble Space Telescope data path.

What is the practical limit of useful magnification for a 150-mm (6-inch) telescope? _____

Answer: 300×.

2.22 TELESCOPE ABERRATIONS

Aberrations are imperfections in the image produced by an optical system.

Chromatic aberration is a lens defect. Starlight consists of all the colors of the spectrum. When starlight passes through a lens, the lens focuses different colors (wavelengths) at slightly different distances. This variation blurs the star image with spurious colors (Figure 2.10). An **achromatic lens**, a combination of two or more lenses made of different kinds of glass, counters this defect.

A properly curved mirror reflects all the colors of starlight to a focus at the same point. The image formed by a reflector has no blurred colors.

Spherical aberration is a mirror defect that blurs a star image. It is a defect of spherical surfaces, hence the name. Parts of the mirror at different distances from the optical axis reflect starlight to slightly different focal points (Figure 2.14).

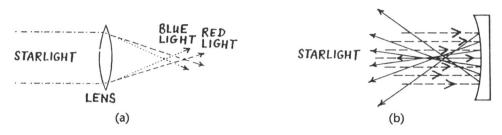

Figure 2.14. (a) Chromatic aberration. A lens bends blue (shorter) light waves the most and brings them to a focus closer to the lens than red (longer) light waves. (b) Spherical aberration. An improperly curved mirror does not reflect light waves to a single focus.

Incorrect curvature of its primary mirror (hyperboloid shape) crippled the Hubble Space Telescope, so its first images were rectified by computer processing. Incorrect curvature can be countered by positioning a correcting lens in front of a mirror so all starlight focuses at the same point. Astronauts repaired the HST's optics and replaced two pairs of gyroscopes and two winglike solar arrays in 1993.

Why should you have the best-quality optical parts in your telescope? _____

Answer: To avoid image aberrations.

2.23 TELESCOPE DESIGN AND SELECTION

You probably wonder which type of telescope is better—a refractor or a reflector. The answer depends on the application involved since each type has advantages and disadvantages over the other.

Small telescopes for astronomy enthusiasts can be of either design. Refractors, with their sealed tubes, are rugged and require less maintenance. But reflectors offer greater aperture for the price and are easier to make at home. The 150-mm (6-inch) Newtonian reflector is popular. Although more expensive per unit of aperture, Schmidt-Cassegrains are the most compact and portable.

Whatever design you choose, the stability of your small telescope **mount** is essential. Nothing will kill your enthusiasm for stargazing faster than a poor-quality telescope with a shaky mount that provides blurry, wiggling images.

Large refractors are used where image quality and resolution are most important, as for viewing surface details of the Moon and planets or for observing double star systems.

Giant reflectors are used where aperture is most important, as to probe the faintest, most distant objects. They are easier to build and are more cost

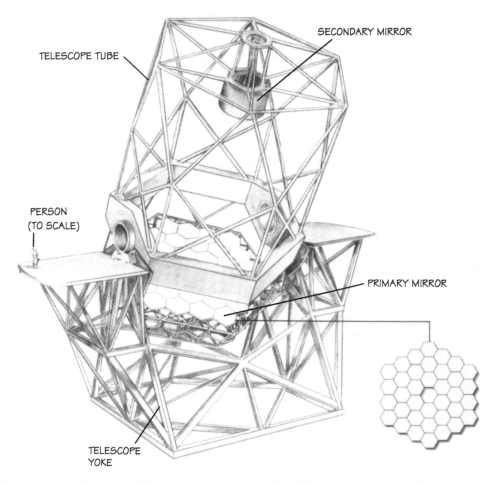

TELESCOPE TUBE

SECONDARY MIRROR

PERSON
(TO SCALE)

PRIMARY MIRROR

TELESCOPE
YOKE

Figure 2.15. The new Keck Telescope in Hawaii, U.S., uses a segmented-mirror design for optical and infrared research. Computer controls precisely align 36 hexagonal mirrors, each about 2 m (6 feet) across and 7.5 cm (3 inches) thick, to form one surface and function as a huge 10-m (33-foot) mirror. Keck I and twin Keck II (under construction nearby) can operate independently or together.

effective than refractors. **Folded optics** reduce the physical length of huge reflectors, so they can be housed inside smaller domes than refractors. The primary mirror is supported from behind so it does not sag under gravity as large lenses do.

Astronomers design ever-larger telescopes and new observing techniques to increase light collection and improve resolving power (Figure 2.15).

The Multiple Mirror Telescope (MMT) is a successful innovation with an electronic computer control system that combines light collection by six 1.8-m mirrors to image like a single 4.5-m mirror equivalent. The MMT is being converted to a single 6.5-m mirror. The largest telescope ever funded is the European Southern Observatory's 16-m Very Large Telescope (VLT), which

TABLE 2.2 Major Optical Telescopes in the World

Project	Size (meters)	Observatory	Location
Very Large Telescope*	16[a]	European Southern	Cerro Paranal, Chile
Keck Telescope 1 and 2*	10	W. W. Keck	Mauna Kea, Hawaii, U.S.
Subaru Telescope*	8.2	National Astronomy Observatory of Japan	Mauna Kea, Hawaii, U.S.
Multiple Mirror Telescope*	6.5[b]	Smithsonian Astrophysical	Mount Hopkins, Arizona, U.S.
Bolshoi Alt-azimuth Telescope	6	Special Astrophysical	Mount Pastukhov, near Zelenchuk-skaya, Russia
George Ellery Hale Telescope	5	Palomar	Palomar Mountain, California, U.S.
William Herschel Telescope	4.2	Royal Greenwich	La Palma, Canary Islands, Spain
4-m Telescope	4	Cerro Tololo Inter-American[c]	Cerro Tololo, Chile
Nicholas U. Mayall Telescope	4	Kitt Peak National[c]	Kitt Peak, Arizona, U.S.
Anglo-Australian Telescope	3.9	Anglo-Australian	Siding Spring Mountain, New South Wales, Australia

*Under construction.

[a] Four separate 8-m (26.2-ft) telescopes.

[b] Conversion.

[c] Facilities of the National Optical Astronomy Observatories (NOAO), headquartered in Tucson, Arizona, U.S.

has a multiple-mirror design utilizing four 8-m telescopes. Astronomers in the U.S., Canada, Europe, and Japan propose to build new, lighter-weight 8-m mirrors to be used both independently and in combination.

Most of the world's biggest telescopes have interesting visitors' centers and self-guided tours for the public (Table 2.2).

Large telescope performance is dramatically improved by new techniques. **Adaptive optics** adjust the mirrors to correct for rapid, hundredths-of-a-second distortions due to turbulence in Earth's atmosphere. **Active optics** correct for minute- or hour-long mirror-shape distortions due to gravity, temperature drifts, and wind.

What are the main advantages of an optical telescope over the unaided eye?

Answer: Superior light-gathering power and resolution. A telescope can also be equipped to record light over a long period of time.

2.24 TELESCOPE ENHANCEMENTS

Research time is in great demand, so astronomers do not sit at giant telescopes and simply stargaze. Instead, observers usually look at a computer display! Starlight, directly or after passing through electronic imaging systems, is recorded for exhaustive study later by many scientists and for obtaining pictures. Powerful **computers** are vital to the acquisition, archiving, processing, and analysis of astronomical data today.

A **charge coupled device (CCD)** is a popular electronic detector. The CCD is a silicon chip of tiny, light-sensitive elements that turns starlight into electric pulses for computers and advanced image processing and display equipment. CCDs are much more sensitive to light than photographic film, and they can record bright and faint objects simultaneously.

Often an instrument called a **spectrograph** is attached to the telescope. Starlight is not a single color but rather a mixture of colors, or wavelengths (Figure 2.16). Astronomers deduce much information about stars from these separate wavelengths, as you will see in Chapter 3.

A **spectroscope** separates starlight into its component wavelengths for viewing. Starlight enters the spectroscope through a narrow slit and goes through a collimating lens, which produces a beam of parallel rays of light. A prism or grating disperses this light into its separate colors (wavelengths). This spectrum is recorded in a spectrograph.

What is the purpose of a spectrograph? _____

Answer: To separate and record the individual wavelengths in a beam of light.

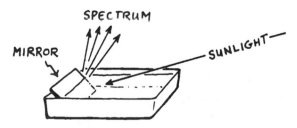

Figure 2.16. You can produce a spectrum from sunlight (starlight). Place a mirror in a pan of water so that it is under the water and leaning against the side of the pan. Position the pan in bright sunlight so that the Sun shines on the mirror. Move the mirror slightly until you see a spectrum on the ceiling or wall.

2.25 RADIO ASTRONOMY

New kinds of telescopes allow today's astronomers to "look" farther into space and "see" more fascinating sights than at any time in the past.

Most **radio telescopes** use a curved "dish" **antenna**, which corresponds to the main mirror in an optical reflector, to collect and focus radio waves from space. This antenna must be very big to collect long radio waves and produce clear images (Figure 2.17).

You cannot see, hear, or photograph these radio waves directly. Instead, they are redirected to a tuned radio receiver that amplifies, detects, and records their electronic image. Computers may display radio images digitized, as a contour map that shows the strength of the radio source (Figure 6.19b) or as a **radiograph** (Figures 6.18 and 6.19a), which is a false color picture that shows how the radio source in space would "look" to a person with "radio vision."

Radio astronomy began in 1931 when U.S. engineer Karl G. Jansky (1905–1950) discovered radio waves coming from the Milky Way. Since then radio waves have been received from diverse sources including our Sun, planets, cold interstellar gas, pulsars, distant galaxies, and quasars.

The world's largest radio antenna is a 300-m (1000-foot) fixed dish built into a valley in the hills at Arecibo, Puerto Rico. The largest fully steerable radio telescope is the 100-m (330-foot) antenna at Effelsberg, Germany.

Figure 2.17. A radio telescope.

The Green Bank Telescope (GBT), under construction at the National Radio Astronomy Observatory (NRAO) in West Virginia, U.S., is to be the most powerful, accurate, and sensitive radio telescope ever built. Its innovative, fully steerable 100-m (330-foot) dish is specially shaped to direct radio waves to the side, where a receiver collects the signals without blocking the dish.

Identify the antenna and prime focus of the radio telescope shown in Figure 2.17. (a) _____ ; (b) _____

Answer: (a) Prime focus; (b) antenna.

2.26 RADIO TELESCOPES

Radio telescopes have several advantages. They let us "see" many celestial objects that emit powerful radio waves but little visible light. They let us "see" radio sources behind interstellar dust clouds in our Milky Way Galaxy that blot out visible stars (because radio waves pass through these clouds). Our atmosphere does not stop or scatter radio waves, so radio telescopes can be used in cloudy weather and during the daytime.

As with optical telescopes, more and clearer radio data can be gleaned by ever-larger collectors. **Aperture synthesis** combines observations from two or more radio telescopes, or **interferometers**, which are linked electronically with computers to achieve the resolving power of one giant collecting dish.

The Very Large Array (VLA) is the primary aperture synthesis facility of the National Radio Astronomy Observatory (Figure 2.18), located at a 2100-m (7000-foot)-high site in New Mexico, U.S. The VLA consists of 27 movable 25-m (82-foot) radio dishes that can be used in different configurations to simulate the performance of a fully steerable radio dish 34 km (21 miles) in diameter. Computers control the antennas, process and display observed data, and produce radiographs with resolution equal to giant optical reflector photographs.

Very long baseline interferometry (VLBI) gives the best resolution. Data are recorded on magnetic tape from coordinated observations of a specific radio source by two or more antennas continents apart. The data can be correlated by computer to simulate one dish as big as Earth.

The U.S. Deep Space Network (DSN) of the National Aeronautics and Space Administration (NASA), maintains radio telescopes on three continents. Stations in the U.S. (California), Spain, and Australia are used for VLBI observations, as well as for space missions. Each has receiving, transmitting, data handling, and interstation communication equipment. The control center is located at the NASA Jet Propulsion Laboratory in Pasadena, California.

The Very Long Baseline Array (VLBA) maps the most distant radio sources and finest details. It consists of 10 automated 25-m (82-foot) radio

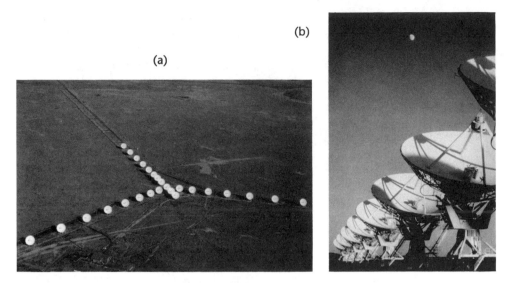

Figure 2.18. (a) The Very Large Array (VLA). (a) Deploys up to 27 antennas in different configurations at 72 observing stations along a Y pattern. Each arm of the Y is about 21 km (13 miles) long. (b) VLA radio telescope.

telescopes distributed across the U.S. from Hawaii to St. Croix, Virgin Islands. Each antenna operates automatically under control from the operations center in New Mexico. Computer processing of the recorded data from all 10 antennas subsequently can synthesize a single radio telescope 8000 km (5000 miles) in diameter.

More resolving power can be achieved by extending VLBI into space.

List at least three advantages of a radio telescope. (1) _____ ;

(2) _____ ;

(3) _____

Answer:

(1) Reveals radio sources—objects that shine in the radio band of wavelengths.

(2) Shows radio sources behind interstellar dust clouds in parts of the Milky Way Galaxy that are hidden from optical viewing.

(3) Works in cloudy weather and during daytime.

(4) Shows radio sources that are located beyond our power of optical viewing.

2.27 INFRARED ASTRONOMY

Infrared telescopes are basically optical reflectors with a special heat detector at the prime focus. Detectors are shielded and cooled to about 2 K to ensure that they register infrared rays from space rather than stray heat from people, equipment, and observatory walls.

Water vapor and carbon dioxide in the air strongly absorb incoming infrared rays. U.S. astronomer Frank Low first built a sensitive infrared detector for astronomy use in 1963. Today large infrared telescopes are located on high mountaintops where the air overhead is thin and dry. The best site is the 4200-m (13,800-foot) summit of Mauna Kea in Hawaii. Smaller telescopes are lofted in airplanes, balloons, rockets, and spacecraft.

The first infrared survey of the whole sky was made in 1983 by the U.S./British/Dutch robot Infrared Astronomical Satellite (IRAS), which identified some 250,000 infrared sources. Astronomers are still analyzing the data.

Infrared telescopes image sources that are relatively cool and often not visible, such as cool stars, dust associated with cold gas and star-forming regions, circumstellar disks (Figure 12.2) with possible extrasolar planets, and comets. Infrared rays pass through interstellar dust more readily than shorter visible rays, revealing the nature of different parts of our Galaxy. Since these rays are not blotted out by sunshine, infrared telescopes can work day and night.

What is the main advantage of infrared telescopes?

Answer: They reveal relatively cool objects that may not be visible.

2.28 ULTRAVIOLET, X-RAY, AND GAMMA RAY ASTRONOMY

High-energy astrophysics is a young, dynamic field in which most discoveries have come since the 1960s. **Ultraviolet, X-ray, and gamma ray telescopes** with suitable detectors are sent above the Earth's obscuring air in orbiting spacecraft. They collect radiation and transmit data to the ground.

The data are processed and recorded by computer for analysis. Data can be displayed digitally or manipulated to generate spectacular **false color images**, in which colors are used to show features of invisible objects.

Ultraviolet observations of the Sun, hot stars, stellar atmospheres, interstellar clouds, a hot gas galactic halo, and extragalactic sources abound. Some of the first came from the still active U.S./European robot International Ultraviolet Explorer (IUE), launched in 1978.

A survey of the sky over the entire extreme ultraviolet (high-energy) spectral region was begun in 1992 by the U.S. robot Extreme Ultraviolet Explorer (EUVE) satellite.

The first survey of the whole sky for X-ray sources was made in 1970 by

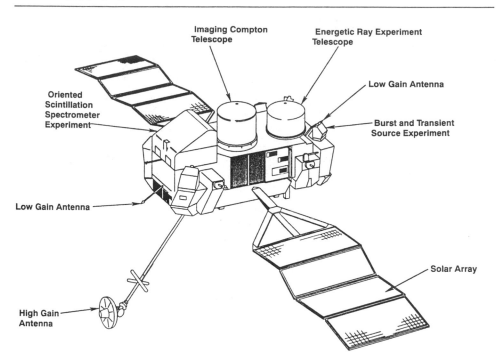

Figure 2.19. Compton Gamma Ray Observatory (GRO) has four instruments to measure gamma rays with the best angular resolution and sensitivity ever.

the U.S. robot Uhuru (Swahili for "freedom"), launched from a European spaceport in Kenya. The German robot Roentgen Satellite (ROSAT), launched by the U.S. in 1990, provided a detailed survey of X-ray sources across the sky followed by studies of the most interesting ones. Astronauts aboard the Earth-orbiting U.S. space shuttle Columbia first operated the reusable observatory Astro-1 in 1990.

Compton Gamma Ray Observatory (GRO), orbited in 1991 by the U.S., images the most energetic objects and violent events in the universe. (Figure 2.19).

Myriad X-ray and gamma ray sources reveal themselves as sudden, intense bursts of radiation (bursters), pulsars, possible black holes, active galaxies, and distant quasars.

What is particularly interesting about new observations by ultraviolet, X-ray, and gamma ray telescopes? _____

_____

_____

Answer: Incoming ultraviolet, X-rays, and gamma rays have much more energy than visible light. They must be generated in extraordinarily energetic processes not yet fully comprehended.

SELF-TEST

This self-test is designed to show you whether or not you have mastered the material in Chapter 2. Answer each question to the best of your ability. Correct answers and review instructions are given at the end of the test.

1. Explain why looking at stars is a way of seeing how the universe looked many years ago. _____

   _____

   _____

   _____

   _____

2. (a) List the major regions of the electronic spectrum from shortest wavelength (highest energy) to longest wavelength (lowest energy). _____

   _____

   _____

   _____

   (b) State what all electromagnetic waves have in common._____

   _____

3. Write the general formula that relates the wavelength and frequency of a wave. _____

4. Suppose you observe a bluish star and a reddish star. State which is hotter, and explain how you know._____

   _____

   _____

5. List the three windows (spectral ranges) in Earth's atmosphere for observational astronomy._____

6. What are the two main parts of a telescope used for stargazing, and what is the function of each? _____

   _____

7. What are the two main advantages of giant telescopes for research?

_____

Two Telescopes

| | Type of Telescope | |
Characteristic	Reflector (1)	Refractor (2)
Diameter of main lens or mirror	2 m	1 m
Focal length of objective	7.6 m	14.6 m
Focal length of eyepiece	5 cm	1 cm

8. Which telescope described in the chart above (1 or 2) has:

(a) greater light-gathering power?

(b) greater resolving power?

(c) greater magnification?

9. What two factors are most important in telescope performance? _____

_____

10. What is the purpose of a spectrograph? _____

_____

11. List three advantages of a radio telescope. _____

_____

_____

12. What is the advantage of sending telescopes up in spacecraft? _____

_____

_____

_____

13. Match an appropriate innovative tool to the observations. _____

_____ (a) Faintest and most distant radio sources.

_____ (b) Very hot stars and gas.

_____ (c) Visible and relatively cool sources.

_____ (d) X-ray sources.

(1) Roentgen Satellite (ROSAT).

(2) Extreme Ultraviolet Explorer (EUVE).

(3) Keck Telescope.

(4) Very Long Baseline Array (VLBA).

ANSWERS

Compare your answers to the questions on the self-test with the answers given below. If all of your answers are correct, you are ready to go on to the next chapter. If you missed any questions, review the sections indicated in parentheses following the answer. If you missed several questions, you should probably reread the entire chapter carefully.

1. Starlight is radiated by electric charges in stars. Light waves transport energy from stars to electric charges in our eyes. Light waves travel incredibly fast—about 300,000 km (186,000 miles) per second. But trillions of miles separate the stars from Earth, and the journey takes many years. Thus we see the stars as they were many years ago when the starlight began its journey to Earth. (Sections 2.1, 2.5)

2. (a) Gamma rays, X-rays, ultraviolet radiation, visible light, infrared radiation, radio waves. (b) All electromagnetic waves travel through empty space at the same speed, the speed of light—about 300,000 km (186,000 miles) per second. (Sections 2.3, 2.5, 2.8)

3.
$$c = f\lambda \text{ or Wavelength} = \frac{\text{Speed of wave}}{\text{Frequency}}$$

(Sections 2.2, 2.5, 2.6, 2.8, 2.9)

4. The bluish star is hotter. The shorter the wavelength at which a star emits its maximum light, the hotter the star, according to Wien's law of radiation. Blue light has a shorter wavelength than red light. (Sections 2.2, 2.10)

5. Optical (visible light), radio, infrared. (Section 2.11)

6. (1) Main mirror or lens (objective): To gather light and form an image. (2) Eyepiece: To magnify the image formed by the main mirror or lens. (Sections 2.12, 2.14, 2.15)

7. Superior light-gathering power and resolving power. (Sections 2.12, 2.19, 2.23)

8. (a) 1; (b) 1; (c) 2. (Sections 2.12, 2.19, 2.20)

9. Size and quality of main mirror or lens. (A stable mount is essential.) (Sections 2.12 and 2.17 through 2.23)

10. To separate and record the individual wavelengths in a beam of light. (Section 2.24)

11. Reveals radio sources; shows radio sources that are hidden from sight behind interstellar dust clouds in the Milky Way Galaxy; works in cloudy weather and daytime; shows radio sources that are located beyond our power of optical viewing. (Sections 2.25, 2.26)

12. Spacecraft take the telescopes beyond Earth's obscuring atmosphere, and it is possible to observe gamma rays, X-rays, and ultraviolet sources that cannot be observed on the ground. There is no atmospheric blurring or radio interference, so a space telescope can work at its practical limit of resolving power.   (Sections 2.11, 2.21, 2.26, 2.27, 2.28)

13. (a) 4;   (b) 2;   (c) 3;   (d) 1.   (Sections 2.23, 2.26, 2.28)

3

THE STARS

Look at the stars! look, look up at the skies!
O look at all the fire-folk sitting in the air!

Gerard Manley Hopkins (1844–1899)
"The Starlight Night"

Objectives

☆ Describe the method and range of the parallax technique for determining the distances to stars.

☆ Describe three types of spectra: emission, absorption, and continuous spectra.

☆ Explain why emission and absorption spectra are unique for each element.

☆ Give a general description of stellar spectra, and explain how they are divided into spectral classes.

☆ Explain how a star's chemical composition, surface temperature, and radial velocity are determined from its spectrum.

☆ List several other kinds of information that are obtained from stellar spectra.

☆ Explain how a star's proper motion and space velocity are determined.

☆ Explain the difference between apparent brightness and luminosity.

☆ Explain the relationship between apparent magnitude, absolute magnitude, and distance.

☆ Describe the H–R diagram and explain the relationship of a star's mass to its luminosity and temperature.

☆ Compare red giants and white dwarfs with our Sun in terms of mass, diameter, and density.

☆ Define four types of binary star systems.

3.1 DISTANCES TO NEARBY STARS

The huge fiery, stars are really trillions of kilometers beyond our atmosphere. The difficult problem of ascertaining the actual distances to the stars has challenged astronomers for centuries.

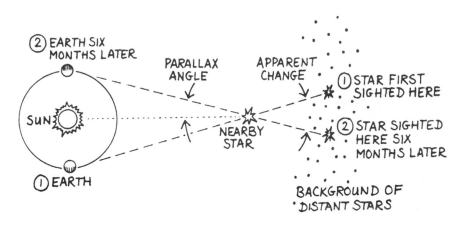

Figure 3.1. Stellar parallax. A nearby star that is sighted from opposite sides of Earth's orbit appears to shift its position from 1 to 2 against a background of distant stars. (The actual parallax angle is extremely tiny.)

The method of **parallax** is used in measuring the distances to nearby stars. The position of a star is carefully determined relative to other stars. Six months later, when Earth's revolution has carried telescopes halfway around the Sun, the star's position is measured again.

Nearby stars appear to shift back and forth relative to more distant stars as Earth revolves around the Sun. The apparent change in a star's position observed when the star is sighted from opposite sides of Earth's orbit is called **stellar parallax**. The distance to the star is calculated from its **parallax angle**, which is one half of the apparent change in the star's angular position. (Figure 3.1).

Stellar parallaxes are very small and are measured in **seconds of arc** (″), where 1″ = 1/3600°. An aspirin tablet would appear to have a diameter of 1″ if it were viewed from a distance of about 2 km (a mile)! The parallaxes of even the nearest stars are less than 1″ (Appendix 5).

One **parsec (pc)**, is the distance to an imaginary star whose *par*allax is 1 *sec*ond of arc (1″). One parsec equals about 31 trillion km (19 trillion miles), or 3.26 light-years.

To calculate the distance to any star from its measured parallax, use the formula:

$$\text{Star's distance (in pc)} = \frac{1}{\text{parallax (″)}}$$

Stellar parallax decreases with the distance of a star. Stellar parallaxes can be measured down to about 0″.01, corresponding to a distance of 100 pc. However, the parallaxes of only a small fraction of the more than 500,000 stars within this distance have been accurately measured.

The European High Precision Parallax Collecting Satellite, or Hipparcos, an **astrometry** spacecraft (1989–1993) measured the positions, parallaxes and motions of 100,000 stars precisely and of another 400,000 stars less exactly. Its

Figure 3.2. Using the parallax method to determine the distances to our closest bright neighbor stars.

name honors Hipparchus (Section 1.7), who calculated the Moon's distance from Earth in 120 B.C. by measuring the Moon's parallax. The *Hipparcos Star Catalog* of star data is being compiled as the voluminous data is processed.

Other indirect methods must be used to determine the distances to the great majority of stars beyond 100 pc.

Would you like to know what "close" means for a star? Refer to Figure 3.2. If the measured parallax for Alpha Centauri is 0".75, then its distance from Earth is 1.3 pc, or 4.3 light-years, which is about 40 trillion km (25 trillion miles). (Alpha Centauri is actually a double star and, with faint Proxima Centauri—the night star closest to us—a member of a triple star system.)

If the measured parallax of Sirius is 0".38, what is its distance from Earth in (a) parsecs? _____ (b) light-years? _____ (c) kilometers or miles (approximate)? _____

Answer: (a) 2.6 pc; (b) 8.5 ly; (c) 81 trillion km or 50 trillion miles.

Solution:

$$(a) \ \frac{1}{0".38}; \quad (b) \ (2.6 \text{ pc}) \times 3.26 \ \frac{\text{ly}}{\text{pc}};$$

$$(c) \ (2.6 \text{ pc}) \times 31 \text{ trillion} \frac{\text{km}}{\text{pc}}$$

$$(2.6 \text{ pc}) \times 19 \text{ trillion} \ \frac{\text{miles}}{\text{pc}}$$

3.2 TYPES OF SPECTRA

Despite the vast distances that separate us from the stars, we know a lot about them. Astronomers can extract an amazing amount of information from starlight.

Remember that starlight is composed of many different wavelengths. When starlight is separated into its component wavelengths, the resulting spectrum holds many clues about the stars. **Spectroscopy** is the analysis of *spectra* (or *spectrums*). Spectra are of three basic types, each produced under different physical conditions.

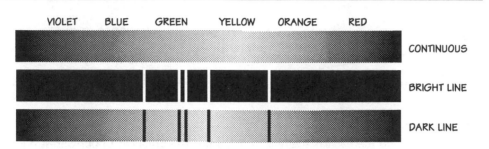

Figure 3.3. The three basic types of spectra as viewed through a spectroscope.

Describe the appearance of each type of spectrum illustrated in Figure 3.3.

(a) _____

(b) _____

(c) _____

Answer:

(a) **Continuous spectrum**: a continuous array of all the rainbow colors.

(b) **Emission**, or **bright-line, spectrum**: a pattern of bright-colored lines of different wavelengths.

(c) **Absorption**, or **dark-line, spectrum**: a pattern of dark lines across a continuous spectrum.

Note: Modern astronomers work with spectra as graphs of intensity versus wavelength (Figure 6.22b).

3.3 SPECTRAL LINES

Atoms are responsible for each type of spectrum. An **atom** is the smallest particle of a chemical element.

More than 100 chemical elements have been identified (Appendix 4). Each element has its own particular kind of atom, first simply described by Danish physicist Niels Bohr (1885–1962).

In the **Bohr atom model**, each element's atoms have a nucleus with a unique number of positively charged protons, circled by the same number of electrons bearing a corresponding negative charge. Atoms are normally electrically neutral.

The electrons are confined to a set of allowed orbits of definite radius. An electron in a particular orbit has a definite **binding energy**, the energy required to remove it from the atom. Each element has its own unique set of allowed electron orbits, or **energy levels**.

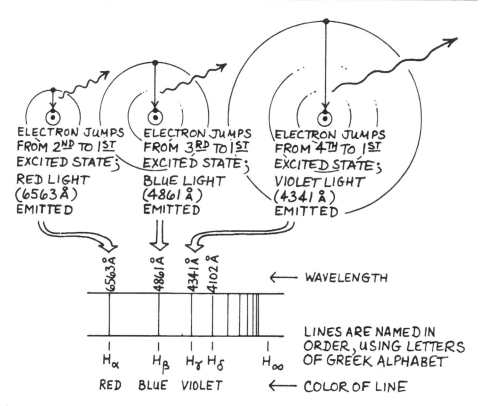

ELECTRON JUMPS FROM 2ND TO 1ST EXCITED STATE; RED LIGHT (6563 Å) EMITTED

ELECTRON JUMPS FROM 3RD TO 1ST EXCITED STATE; BLUE LIGHT (4861 Å) EMITTED

ELECTRON JUMPS FROM 4TH TO 1ST EXCITED STATE; VIOLET LIGHT (4341 Å) EMITTED

6563 Å   4861 Å   4341 Å   4102 Å   ← WAVELENGTH

H_α H_β H_γ H_δ H_∞

RED   BLUE   VIOLET

LINES ARE NAMED IN ORDER, USING LETTERS OF GREEK ALPHABET

← COLOR OF LINE

Figure 3.4. Origin of the unique set of bright red, blue, and violet emission lines of hydrogen.

An undisturbed atom, in the **ground state**, has the least possible energy. If the right energy is supplied, an electron will jump to a higher energy level. Then the atom is in an unstable **excited state**. When the electron falls back down, the atom radiates that energy in the form of a pellet of light called a **photon**.

If an atom absorbs enough energy, one or more of its electrons can be removed completely. The atom, which is left with an electric charge (positive), is then called an **ion**.

Bright-colored **emission lines** are produced when electrons jump from higher energy levels back down to lower energy levels. The wavelength of the light emitted is inversely proportional to the energy difference between the energy levels. Since each kind of neutral or ionized atom has its own unique set of energy levels, each chemical element has its own unique set of bright-colored emission lines (Figure 3.4).

Corresponding unique **dark absorption lines** are produced when an atom of a chemical element absorbs light and the electrons jump out to higher energy levels (Figure 3.5).

Thus, an emission spectrum or a corresponding absorption spectrum gives positive identification of the chemical element that produced it.

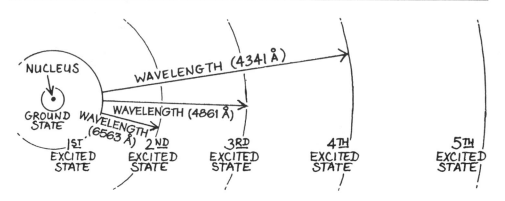

Figure 3.5. Origin of the dark absorption lines corresponding to the bright red, blue, and violet emission lines of hydrogen.

Why do atoms emit light of different colors (specific wavelengths)? _____

_____

_____

Answer: Each color (wavelength) corresponds to an electron jumping down from a particular higher energy level to a particular lower energy level.

3.4 SPECTRA OF STARS

Stellar spectra, or the spectrums of stars, are predominantly patterns of dark lines crossing a continuous band of colors (Figure 3.8). An amateur U.S. astronomer, Henry Draper (1837–1882), first photographed the spectrum of a star in 1872.

Stars are blazing balls of gas where many kinds of atoms emit light of all colors. Light from a star's bright, visible surface, called its **photosphere**, is blurred into a continuous spectrum of colors. As the light travels through the star's outer atmosphere, some of the colors (photons of certain wavelengths) are absorbed, producing dark absorption lines. These absorption lines identify the chemical elements that make up the star's atmosphere.

Refer to Figure 3.6. Identify the region of a star where (a) a continuous spectrum and (b) an absorption spectrum would originate. (a) _____ ;
(b) _____

Answer: (a) Continuous; (b) absorption.

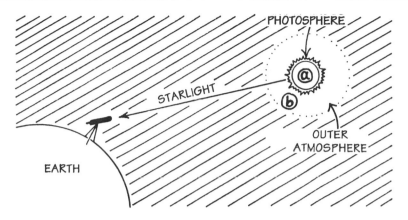

Figure 3.6. The starlight we observe comes from a star's photosphere and passes through its outer atmosphere before shining into space.

3.5 CHEMICAL COMPOSITION

Our Sun was the first star whose absorption spectrum was analyzed. In 1814, Bavarian physicist Joseph von Fraunhofer (1787–1826) recorded the strongest dark lines, now called **Fraunhofer lines**.

Since then astronomers have catalogued thousands of dark lines in the Sun's spectrum. By comparing these lines with the spectral lines produced by different chemical elements on Earth, they have found over 70 different chemical elements in the Sun (Figure 3.7).

How can the chemical composition of stars be determined? Assume that stars and their atmospheres are made of the same ingredients. _____

Answer: By analyzing the dark lines in the star's spectrum and comparing them with those of each of the chemical elements on Earth.

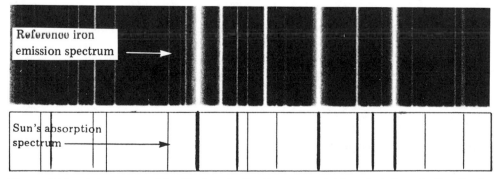

Figure 3.7. Astronomers identify iron in the Sun by matching dark lines in the Sun's absorption spectrum to a reference iron emission spectrum.

3.6 SPECTRAL CLASSES

When you compare the spectra of stars like Polaris or Vega with the Sun's spectrum (Figure 3.8), you see that some look the same while others look quite different. Absorption spectra are used to classify stars into seven principal types, called **spectral classes**.

Hydrogen lines are much stronger in the spectra of some stars than in the Sun's spectrum. Astronomers once mistakenly thought that these stars had more hydrogen than other stars. They classified stars by the strength of the hydrogen lines in their spectra, in alphabetical order, from the strongest (called Class A) to the weakest (called Class Q).

U.S. astronomer Annie J. Cannon (1863–1941), who examined and classified the spectra of 225,300 stars, modified this classification system to its pre-

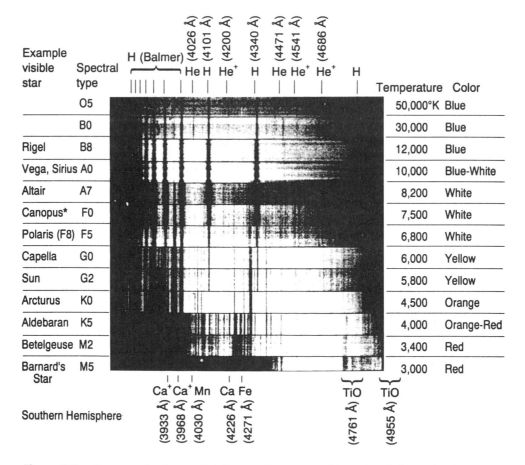

Figure 3.8. Seven main classes of stellar spectra, arranged in order of decreasing temperature. The wavelengths of spectral lines are given in angstrom units (Å). (He: neutral helium; H: hydrogen; Ca: calcium; Fe: iron; TIO: titanium oxide; He$^+$: helium ion; Ca$^+$: calcium ion.)

sent form as O B A F G K M. (Astronomy students remember this strange order by saying: "*Oh Be A Fine Girl/Guy Kiss Me*.")

Today we know that all visible stars are roughly uniform in composition. All are made mostly of hydrogen and helium. U.S. astronomer Cecilia Payne-Gaposhkin (1900–1979) demonstrated that the differences in the dark line patterns of stars are due primarily to their enormously different **surface temperatures**.

Now the sequence of spectral classes identified by the traditional letters is recognized as a temperature sequence. The O stars are hottest, with the temperature continuously decreasing down to the coolest M stars. Each spectral class is arranged in 10 subclasses numbered 0 to 9, also in order of decreasing temperature.

What property determines the spectral class of a star? _____

Answer: Surface temperature.

3.7 TEMPERATURE

The spectrum of a hot star and that of a cool star look very different. Examine Figure 3.8. The photographs represent the seven main classes of stellar spectra. Each spectral class has characteristics that serve, like numbers on a thermometer, to indicate a star's temperature.

The spectral classes of stars in order from highest to lowest temperatures, the approximate surface temperatures of these classes, and the main class characteristics are summarized in Table 3.1.

You can identify a new star's spectral class and probable temperature by comparing its spectrum to the photographs in Figure 3.8 and the class characteristics in Table 3.1.

TABLE 3.1   Spectral Class Characteristics

Spectral Class	Approximate Temperature (K)	Main Class Characteristics
O	>30,000	Relatively few lines; lines of ionized helium
B	10,000–30,000	Lines of neutral helium
A	7,500–10,000	Very strong hydrogen lines
F	6,000–7,500	Strong hydrogen lines; ionized calcium lines; many metal lines
G	5,000–6,000	Strong ionized calcium lines; many strong lines of ionized and neutral iron and other metals
K	3,500–5,000	Strong lines of neutral metals
M	<3,500	Bands of titanium oxide molecules

Lines of ionized helium

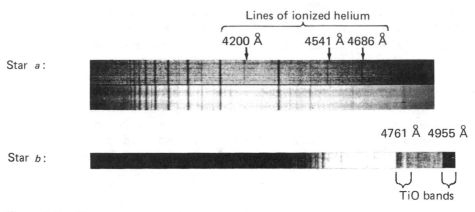

4200 Å   4541 Å 4686 Å

Star *a*:

4761 Å 4955 Å

Star *b*:

TiO bands

Figure 3.9.   Star spectra.

List the spectral class and probable temperature of each of the stars whose spectrum is shown in Figure 3.9. (a) _____ ; (b) _____

Answer: (a) O type (>30,000 K); (b) M type (<3500 K).

3.8 ORIGIN OF SPECTRAL CLASS CHARACTERISTICS

Atomic theory explains why hot blue (O-type) stars and cool red (M-type) stars produce spectra that look so different even though all stars are made of practically the same ingredients.

Every chemical element has a characteristic temperature and density at which it is most effective in producing visible absorption lines.

At extremely high temperatures, as in O stars, gas atoms are **ionized**, or broken up. Only the most tightly bound atoms such as singly ionized helium survive, and the lines of ionized atoms dominate the spectrum. When the temperature is around 5800 K, as in G stars such as our Sun, metal atoms such as iron and nickel remain neutral without being disrupted. At temperatures below 3500 K, as in M stars, even molecules such as titanium oxide can exist.

Does the absence of the characteristic absorption lines of a particular element like hydrogen in a star's spectrum necessarily mean that the star does not contain that element? Explain._____

_____

Answer: No. The star's temperature determines which kinds of atoms can produce visible absorption lines.

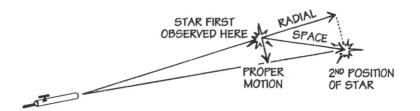

Figure 3.10. Space velocity has two components—radial velocity and proper motion.

3.9 MOTIONS

Stars have a **space velocity**, or motion through space with respect to the Sun, of many kilometers a second.

Space velocity has two components, which are measured independently: **radial velocity**, or speed toward or away from us along the line of sight; and **proper motion**, or the amount of angular change in a star's position per year (Figure 3.10).

A star's radial velocity is determined from analysis of its spectrum. The **Doppler shift** is an effect, discovered by Austrian physicist Christian Doppler (1803–1853), that applies to all wave motion. When a source of waves and an observer are approaching or receding from each other, the observed wavelengths are changed.

A star's spectral lines (wavelengths) for any given element, such as iron, are compared with a reference spectrum. The star's wavelengths are shorter (**blueshift**) or longer (**redshift**) according to whether the star is moving toward or away from us (Figure 3.11).

The change in wavelength ($\Delta\lambda$) divided by the wavelength from a stationary source (λ) is proportional to the relative velocity (v) (unless v is comparable to the velocity of light, c). The formula is:

$$\frac{\Delta\lambda}{\lambda} = \frac{v}{c}$$

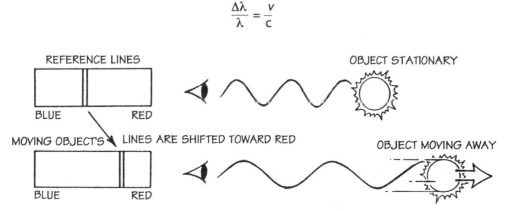

Figure 3.11. Doppler shift. A sky object's spectral lines for any given element are compared with reference lines. Redshifted spectral lines indicate that the object is moving away from us.

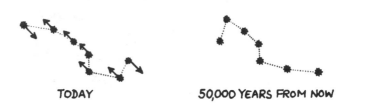

TODAY   50,000 YEARS FROM NOW

Figure 3.12. Measured proper motion of the Big Dipper today indicates that the grouping will have a whole new look far in the future.

Proper motion is measured over an interval of 20 to 30 years. The average proper motion for all visible stars is less than 0.1 second of arc (0".1) per year. At that rate you won't notice any change in the appearance of your favorite constellation during your lifetime. But if you could return to observe the sky 50,000 years from now, it would look very different (Figure 3.12).

What is the angular change in an average visible star's position after 50,000 years? _____

Answer: 5000 seconds of arc, or 1.39° (almost three times the Moon's angular diameter, which is ½°).

Solution: 0".1 per year ✕ 50,000 years = 5000"

3.10 OTHER PROPERTIES

Other information about stars is obtained from careful measurements of **spectral line shape**.

Gas **density**, the mass per unit volume, is indicated by **collisional broadening**. A broadened spectral line is produced when atoms collide more frequently in higher-density stars.

Axial rotation, the rotation of a star around its axis, is indicated by **rotational broadening**. If observable, a broadened line can yield a lower limit to the star's rate of rotation on its axis.

A splitting or broadening of spectral lines occurs in the presence of a **magnetic field**, a region where magnetic forces are detected, which is called the **Zeeman effect**. The amount of splitting depends on the magnetic field strength.

These different kinds of broadening are not distinguishable to the unaided eye but are determined by careful analysis of the shape of the line using sensitive spectrometers.

List three properties of a star connected to its spectral line shape. (1) _____ ; (2) _____ ; (3) _____

Answer: (1) Density; (2) axial rotation; (3) magnetic field strength.

3.11 DECODING A STAR'S SPECTRUM

Write a brief paragraph summarizing your understanding of how astronomers deduce different properties of a star from its spectrum. _____

_____

_____

_____

_____

Answer: Your summary should include the following concepts: (1) chemical composition, from the presence of the characteristic lines of certain elements; (2) temperature, from spectral type; (3) star's speed toward or away from us, from Doppler shift of the lines; (4) density, axial rotation, and surface magnetic fields, from line shape.

3.12 LUMINOSITY

Astronomers distinguish a star's **apparent brightness**—the way the star appears in the sky—from its **luminosity**—the actual amount of light a star shines into space each second.

The star we know best is our Sun. The luminosity of other stars is often stated in terms of the **Sun's luminosity** ($L_\odot$), which is 3.85×10^{26} watts. The Sun's luminosity is equivalent to 3850 billion trillion 100-watt light bulbs shining all together.

The most luminous stars are over a million times as luminous as the Sun. The dimmest stars known are less than 0.0001 as luminous as the Sun.

Rigel in Orion is about 60,000 times more luminous than the Sun.

Explain why the Sun looks much brighter to us than Rigel does.

_____

_____

_____

_____

Answer: Rigel is over 50 million times farther away from Earth (1400 light-years) than the Sun is (about 150 million km, or 93 million miles). A star's apparent brightness depends on both its luminosity and its distance from us.

3.13 PROPAGATION OF LIGHT

You cannot tell by looking at stars in the sky which ones have the greatest luminosity. The farther away a star is, the less bright it appears.

Light spreads out uniformly in all directions from a source so that the amount of starlight shining on a unit area falls off as the square of the distance away from the star. This relationship is called the **inverse square law** (Figure 3.13). Thus, if two stars have exactly the same luminosity but one is twice as far away from you as the other, the distant one will look only $\frac{1}{2^2} = \frac{1}{4}$ as bright as the closer one, because you get one fourth the light in your eyes.

Our Sun is exceptionally bright because it is so close to us. If it were located 100,000 times deeper in space, how many times fainter would it look?

Answer: 10 billion times fainter, or about like the brilliant blue-white star, Sirius.

Solution: $\dfrac{1}{(100,000)^2} = \dfrac{1}{10,000,000,000}$ as bright (or 10 billion times fainter)

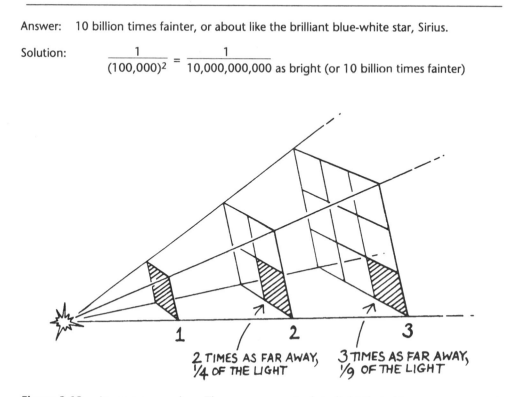

Figure 3.13. Inverse square law. The same amount of starlight that shines on a square at 1 spreads out to illuminate four equal squares at 2 and nine equal squares at 3. So the star's apparent brightness decreases as the (distance)^2 increases.

3.14 APPARENT MAGNITUDE

Apparent magnitude is a measure of how bright a star appears (see Section 1.7). The modern **magnitude scale** defines a first-magnitude star to be exactly 100 times brighter than a sixth-magnitude star.

This ratio agrees with the way our eyes respond to increases in the brightness of stars. What we see as a linear increase in brightness (a difference of one magnitude) is precisely measured as a geometrical increase in brightness (the fifth root of 100 or 2.512 times brighter).

Magnitude differences between stars measure the relative brightness of the stars. Table 3.2 lists approximate brightness ratios corresponding to sample magnitude differences.

Remember that the most negative magnitude numbers identify the brightest objects, while the largest positive magnitude numbers identify the faintest objects.

Refer to Tables 3.2 and 3.3. How much brighter does the Sun appear than Sirius? Explain.

_____

_____

Answer: 10 billion times brighter.

Solution: Magnitude difference is (−26.7) − (−1.5) ≈ 25, corresponding to a brightness ratio of 10,000,000,000:1.

TABLE 3.2 Magnitude Differences and Brightness Ratios

Difference in Magnitude	Brightness Ratio
0.0	1:1
1.0	2.5:1
2.0	6.3:1
3.0	16:1
4.0	40:1
5.0	100:1
6.0	251:1
10.0	10,000:1
15.0	1,000,000:1
20.0	100,000,000:1
25.0	10,000,000,000:1

TABLE 3.3   Sample Magnitude Data

Subject	Description	Apparent Magnitude	Absolute Magnitude
Sun		–26.7	4.8
100-watt bulb	At 3 m (10 ft)	–18.7	66.3
Moon	Full	–12.5	32
Venus	At brightest	–4.7	28
Sirius	Brightest star	–1.5	1.4
Alpha Centauri	Closest seeable star	0	4.4
Andromeda Galaxy	Farthest seeable object	3.5	–21

3.15 ABSOLUTE MAGNITUDE

Absolute magnitude is a measure of luminosity, or how much light a star is actually radiating into space. If you could line up all stars at the same distance from Earth, you could see how they differ in intrinsic, or "true," brightness.

Astronomers define a star's absolute magnitude as the apparent magnitude the star would have if it were located at a standard distance of 10 parsecs from us. With the effects of distance canceled out, they can use absolute magnitude comparisons to determine differences in the actual light output of stars (Figure 3.14).

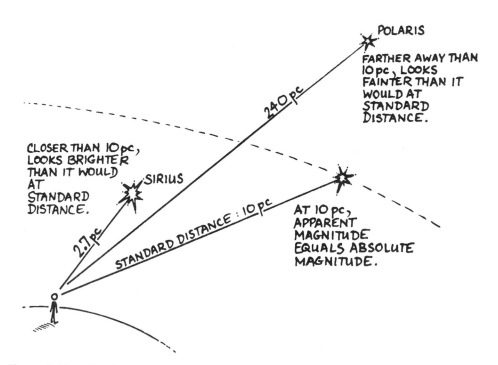

Figure 3.14.   Absolute magnitude and apparent magnitude.

If a star is farther than 10 parsecs from us, its apparent magnitude is numerically bigger than its absolute magnitude. (Large positive magnitude numbers indicate faint objects.) For example, Polaris is 240 pc away. Its apparent magnitude is +2.3, whereas its absolute magnitude is – 4.6.

On the other hand, if a star is closer than 10 parsecs, its apparent magnitude is numerically smaller than its absolute magnitude. Thus, Sirius is 2.76 pc away. Its apparent magnitude is –1.5, whereas its absolute magnitude is only +1.4.

Consider the two bright stars Deneb and Vega. Refer back to Table 1.1 to fill in the chart below. Then tell (a) which looks brighter? _____ (b) Which is really more luminous? _____ (c) What factor makes your answers to (a) and (b) different? _____

Two Bright Stars

Star	Constellation	Apparent Magnitude	Absolute Magnitude
Deneb	Cygnus	(a)	(b)
Vega	Lyra	(c)	(d)

Answer: Chart: (a) 1.25; (b) –7.2; (c) 0.03; (d) 0.6.

(a) Vega (numerically smaller apparent magnitude). (b) Deneb (numerically more negative absolute magnitude). (c) The distance the stars are from us.

3.16 DISTANCES FROM MAGNITUDES

The difference between the apparent magnitude (m) and absolute magnitude (M) is called the **distance modulus (m – M)**. In formula form:

$$m - M = 5 \log \left(\frac{\text{distance in parsecs}}{10} \right)$$

A star's apparent magnitude can be measured directly. For a distant star whose parallax cannot be measured but whose absolute magnitude is known, as from consideration of its spectrum, the distance modulus can be used to calculate distance.

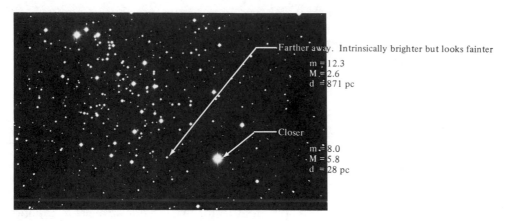

Figure 3.15. Star clusters, showing difference in apparent and absolute magnitude. Farther away (intrinsically brighter but looks fainter): m = 12.3; M = 2.6; d = 871 pc. Closer: m = 8.0; M = 5.8; d = 28 pc.

Refer to Figure 3.15. Give the distance modulus of the stars that are (a) closer _____ ; (b) farther away _____

Answer: (a) 2.2; (b) 9.7.

3.17 COMPARISONS

Be sure you understand the ideas presented so far by answering the following questions about four of the Sun's neighbor stars described in the chart below.

Four Nearby Stars

Star	Apparent Magnitude	Absolute Magnitude	Spectral Class	Parallax (in")
Alpha Centauri	−0.3	4.1	G	0.750
Thuban	4.7	5.9	K	0.176
Barnard's Star	9.5	13.2	M	0.545
Altair	0.8	2.3	A	0.202

Which star is (a) hottest? _____ (b) coolest? _____
(c) brightest looking? _____ (d) faintest appearing? _____
(e) intrinsically (actually) most luminous? _____ (f) intrinsically least
luminous? _____ (g) closest? _____ (h) most
distant? _____ Explain your answers.

Answer: (a) Altair, spectral class A; (b) Barnard's Star, spectral class M; (c) Alpha Centauri, apparent magnitude 0.0; (d) Barnard's Star, apparent magnitude 9.5; (e) Altair, absolute magnitude 2.2; (f) Barnard's Star, absolute magnitude 13.2; (g) Alpha Centauri, parallax = 0".750, or distance = 1/parallax = 1/0".750 = 1.3 pc; (h) Thuban, parallax = 0".176, or distance = 1/0".176 = 5.7 pc.

3.18 HERTZSPRUNG–RUSSELL DIAGRAM

A basic link between luminosities and temperatures of stars was discovered early in the twentieth century by two independent astronomers, Henry N. Russell (1877–1957) of the U.S. and Ejnar Hertzsprung (1893–1967) of Denmark. The **Hertzsprung–Russell (H–R) diagram** is a plot of luminosity versus temperature. Astronomers use the H–R diagram widely to check their theories (Figure 3.16).

Every dot on an H–R diagram represents a star whose temperature (spectral class) is read on the horizontal axis and whose luminosity (absolute magnitude) is read on the vertical axis.

Significantly, when a few thousand stars are chosen randomly and plotted on an H–R diagram, they fall into definite regions. This pattern indicates that

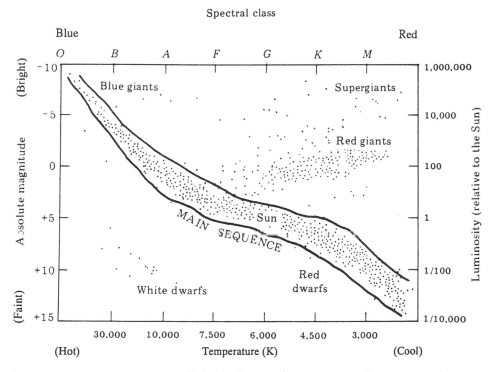

Figure 3.16.   Hertzsprung–Russell (H–R) diagram for many stars. Temperature increases from right to left. Luminosity increases from bottom to top.

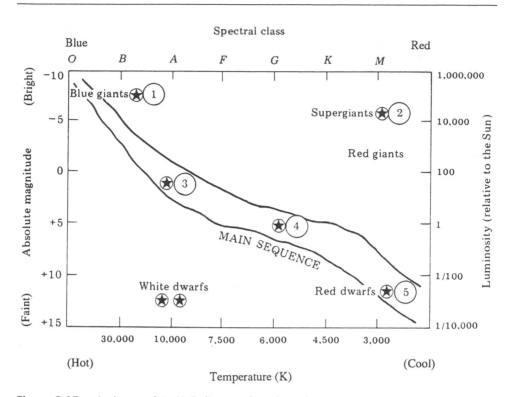

Figure 3.17. An incomplete H–R diagram for selected stars.

a meaningful connection exists between a star's luminosity and its temperature. Otherwise, the dots would be scattered randomly all over the graph.

About 90 percent of the stars lie along a band called the **main sequence**, which runs from the upper left (hot, very luminous **blue giants**) across the diagram to the lower right (cool, faint **red dwarfs**). Red dwarfs are the most common type of nearby star.

Most of the other 10 percent of stars fall into the upper right region (cool, bright **giants** and **supergiants**) or in the lower left corner (hot, low-luminosity **white dwarfs**).

Identify the location of the following stars indicated on the H–R diagram in Figure 3.17. Each star's absolute magnitude is given in parentheses. Refer to Figure 3.8 for temperature and spectral class. (a) Rigel (–8.1) _____ ; (b) Vega (0.6) _____ ; (c) Sun (4.8) _____ ; (d) Betelgeuse (–7.2) _____ ; (e) Barnard's Star (13.2) _____

Answer: (a) 1; (b) 3; (c) 4; (d) 2; (e) 5.

3.19 MASS-LUMINOSITY RELATION

A star's position on the main sequence is determined by its **mass**, or the amount of matter the star contains.

The main sequence is a sequence of stars of decreasing mass, from the most massive, most luminous stars at the upper end to the least massive, least luminous stars at the lower end (Figure 3.18).

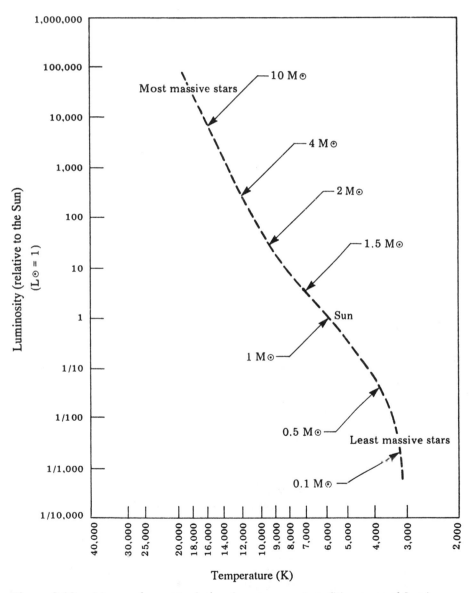

Figure 3.18. Masses of some typical main sequence stars. ($M_⊙$ = mass of Sun.)

An empirical **mass–luminosity relation** for main sequence stars, found from binary stars, says that the more massive a star is, the more luminous it is. The luminosity of a star is approximately proportional to its mass raised to the 3.5 power.

The **mass of the Sun, $M_\odot$** = 2×10^{30} kg, is practically 333,000 times the mass of Earth. Stellar masses do not vary enormously along the main sequence as stellar luminosities do. The faintest red dwarfs have a mass about one tenth of the Sun's. (A hypothetical object with a mass between ¹⁄₁₀₀ and ¹⁄₁₀ the Sun's, called a **brown dwarf**, may shine briefly but is too small to get hot enough to become a star.) The largest mass of a stable star is about 60 to 75 times that of the Sun.

What basic property of a star determines its position on the main sequence of the H–R diagram; that is, what determines its luminosity and temperature?

Answer: Mass.

3.20 SIZES AND DENSITIES ⭐

Our Sun is the only star that is close enough to allow astronomers to measure its size directly.

The diameter of the Sun is 1.39 million km (about 864,000 miles). That is equal to 109 Earths placed next to each other.

If the absolute temperature and luminosity of a star are determined, then the size of the star can be calculated from the **Stefan–Boltzman radiation law**. This law says that the luminosity (L) of a star is proportional to the square of its radius (R) times the fourth power of its surface temperature (T). The equation is:

$$L = 4\pi R^2 \sigma T^4$$

where σ is the Stefan–Boltzman constant (Appendix 2).

Stars on the main sequence vary in size continuously from the large blue-white giants, about 25 times the Sun's *radius* ($R_\odot$), down to the common cool red dwarfs, which are only about ¹⁄₁₀ the Sun's radius.

The largest stars are the supergiants such as Betelgeuse in Orion, whose radius is about 400 times the Sun's radius. You could fit more than a million stars like our Sun inside Betelgeuse! The smallest common stars are the white dwarfs, roughly the size of Earth.

The mean **density**, or the mass per unit volume, of the Sun is 1.4 g/cm³, slightly more than that of water. Red giant stars and white dwarf stars both have about the same mass as the Sun but are very different sizes.

What can you say about the densities of red giants and white dwarfs compared to the Sun? Explain. _____

_____

_____

_____

_____

_____

Answer: Red giants have very low density compared to the Sun. They have the same amount of mass in a much bigger volume. (Their mean density is about the same as that of a vacuum here on Earth.)

White dwarfs are extremely dense. They have the same amount of mass packed into a much smaller volume. (One teaspoon of material from a white dwarf would weigh several tons on Earth.)

3.21 DOUBLE STAR SYSTEMS

Many stars that look single to the unaided eye are not. A **binary star** is formed by a pair of stars that revolve around a common center of gravity as they travel through space together (Figure 3.19). The masses of the stars can be figured from the angular size and period of their orbits.

Binary stars are classified by the way they are observed.

A **visual binary** can be resolved with a telescope so that two separate stars can be seen. Over 70,000 visual binaries are known. Mizar in Ursa Major

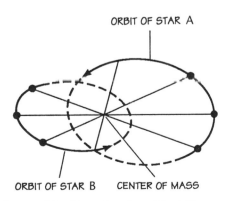

Figure 3.19.   A binary star consists of two stars A and B orbiting a common center of mass, bound together by their mutual gravity.

was the first binary star discovered, in 1650. Beautiful Albireo in Cygnus is a colorful yellow and blue star. You can see these and many others in a small telescope. (See "Useful Resources" for observers' guides).

Many visible stars may have companions that are too faint to be seen. An **astrometric binary** is a visible star plus an unseen companion star. The presence of an unseen companion is inferred from variable proper motion of the visible star. Brilliant Sirius (Sirius A) in Canis Major was an astrometric binary from 1844, when its nature was detected, until 1862. Then its faint companion star (Sirius B) was observed.

A **spectroscopic binary** cannot be resolved in a telescope. Its binary nature is revealed by its spectrum. A varying Doppler shift is apparent in the spectral lines as the stars approach and recede from Earth. Almost a thousand spectroscopic binaries have been analyzed. The brighter member of Mizar (Mizar A) is a spectroscopic binary.

An **eclipsing binary** is situated so that one star passes in front of its companion, cutting off light from our view at regular intervals. An eclipsing binary regularly changes in brightness. You can see the famous eclipsing binary Algol, the Demon, in Perseus. Algol "winks" from brightest magnitude 2.2 to least bright magnitude 3.5 in about 2 days and 21 hours.

An **optical double** is a pair of stars that appear to be close to each other in the sky when viewed from Earth. Actually, one is much more distant than the other, and they have no physical relationship to one another.

Test your eyesight by finding both Mizar and Alcor (nicknamed "the testers"), the optical double in the handle of the Big Dipper in Ursa Major.

How does an optical double differ from a visual binary?_____

_____

_____

Answer: The stars in an optical double are far apart and have no actual relationship to one another. The stars in a visual binary are bound together in space by their mutual gravity.

SELF-TEST

This self-test is designed to show you whether or not you have mastered the material in Chapter 3. Answer each question to the best of your ability. Correct answers and review instructions are given at the end of the test.

1. Refer to the chart on page 78. From the measured parallax, find the distance to Barnard's Star in (a) parsecs _____ ; (b) light-years _____ .

2. Explain why the bright (dark) spectral lines of light emitted from (absorbed by) the atoms of an element are unique to that element. _____

   _____

   _____

   _____

   _____

3. Explain how a star's spectrum is formed. _____

   _____

   _____

   _____

   _____

4. List the following types of spectral lines in order as they appear in stars of decreasing temperature.

   ____ (1) Very strong hydrogen lines.   ____ (4) Neutral helium.
   ____ (2) Ionized helium.   ____ (5) Neutral metals.
   ____ (3) Bands of titanium oxide   ____ (6) Ionized metals.
             molecules.

5. Match the following properties of a star that can be deduced from its spectrum with the appropriate method listed on the right.

   ____ (a) Chemical composition.   (1) Doppler shift.
   ____ (b) Temperature.   (2) Spectral type (class).
   ____ (c) Radial velocity.   (3) Line shape.
   ____ (d) Gas density, axial rotation, (4) Characteristic lines.
             magnetic field.

6. The proper motion of Sirius is 1.32″ per year. Find how much Sirius will change its position on the celestial sphere in the next 1000 years. _____

_____

7. Define space velocity. _____

_____

8. Refer to Table 1.1. By using their apparent magnitudes, absolute magnitudes, and spectral classes, match one of the four stars to each description.

   ____ (a) Hottest.   (1) Betelgeuse.
   ____ (b) Coolest.   (2) Procyon.
   ____ (c) Most luminous.   (3) Spica.
   ____ (d) Least luminous.   (4) Sirius.
   ____ (e) Brightest.
   ____ (f) Faintest.
   ____ (g) Closest.
   ____ (h) Most distant.

9. Label the following on the H–R diagram in Figure 3.20:
   (1) Surface temperature of star (K).   (6)   Red giants.
   (2) Absolute luminosity (Sun = 1).   (7)   White dwarfs.
   (3) Spectral class.   (8)   Supergiants.
   (4) Absolute magnitude.   (9)   Blue giants.
   (5) Main sequence.   (10) Red dwarfs.

10. What is the most basic property of a star that determines its location on the main sequence (its temperature and luminosity)? _____

11. Use the H–R diagram to explain why, compared to our Sun, red giants must be very large and white dwarfs must be very small. _____

_____

_____

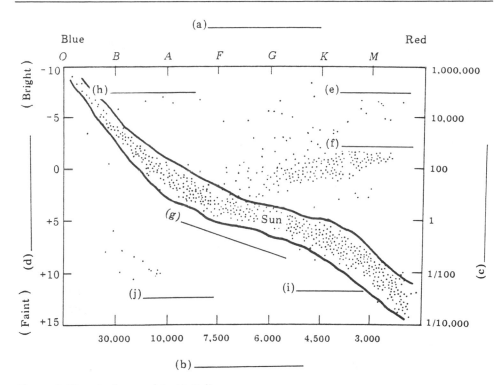

Figure 3.20. An incomplete H–R diagram.

12. Match:

____ (a) Can be resolved with a telescope.

____ (b) Unseen companion inferred from variable proper motion of visible companion.

____ (c) Binary nature revealed by its spectrum.

____ (d) Changes in brightness regularly as one star blocks its companion from our view.

____ (e) Member stars have no actual physical relationship to one another.

(1) Astrometric binary.
(2) Eclipsing binary.
(3) Optical double.
(4) Spectroscopic binary.
(5) Visual binary.

Compare your answers to the questions on the self-test with the answers given below. If all of your answers are correct, you are ready to go on to the next chapter. If you missed any questions, review the sections indicated in parentheses following the answer. If you missed several questions, you should probably reread the entire chapter carefully.

1. (a) 1.8 pc;   (b) 6.0 ly.   (Section 3.1)

   Solution: Measured parallax is

   $$0''.545 \text{ and } \frac{1}{0''.545} = 1.8 \text{ pc}$$

2. Each spectral line is light of a particular wavelength emitted (or absorbed) by the atom when one of its electrons jumps between a higher and a lower energy level (orbit). Each element has its own unique set of allowed electron orbits, so each element has its own characteristic set of spectral lines. (Sections 3.2, 3.3)

3. Stars are blazing balls of gas where many kinds of atoms emit light of all colors. This light, emitted from the star's surface, passes through the star's outer atmosphere. There, atoms of each element absorb their characteristic wavelengths, so a pattern of dark lines crosses the continuous band of colors—the star's spectrum.   (Sections 3.3, 3.4)

4. 2;   4;   1;   6;   5;   3.   (Section 3.7)

5. (a) 4;   (b) 2;   (c) 1;   (d) 3.   (Sections 3.3, 3.5 through 3.10)

6. 1320", or roughly one third of a degree.

   Solution: Proper motion = 1.32" per year. A degree is equal to 3600".

   (1.32" per year) × 1000 years   (Section 3.9)

7. Velocity of a star with respect to the Sun.   (Section 3.9)

8. (a) 3;   (b) 1;   (c) 1;   (d) 2;   (e) 4;   (f) 3;   (g) 4;   (h) 1. (Sections 3.6, 3.7, 3.12 through 3.16)

Solution:

	Spectral Class	Distance (ly)	Apparent Magnitude	Absolute Magnitude
Betelgeuse	M	1400	0.50	−7.2
Procyon	F	11.4	0.38	2.6
Spica	B	220	0.98	−3.2
Sirius	A	8.6	−1.46	1.4

9. (a) 3;   (b) 1;   (c) 2;   (d) 4;   (e) 8;   (f) 6;   (g) 5;   (h) 9;

   (i) 10;   (j) 7.   (Section 3.18)

10. Mass.   (Section 3.19)

11. Red giants are relatively cool but luminous; hence, they must have a large surface area radiating energy. White dwarfs are relatively hot but faint; hence, they must have a small surface area radiating energy into space. (Sections 3.18 through 3.20)

12. (a) 5;   (b) 1;   (c) 4;   (d) 2;   (e) 3.   (Section 3.21)

4

THE SUN

*Thou dawnest beautifully in the horizon of the sky
O living Aton who wast the Beginning of life!*

Akhenaton (c. 1386–1358 B.C.)
"Hymn to the Sun"

Objectives

☆ List some reasons why modern astronomers study the Sun.

☆ Define the solar constant, and explain why it is important to know if it is truly constant with time.

☆ Define the astronomical unit, AU.

☆ Relate the formation, properties, and motions of the Sun as a star.

☆ Sketch the structure of the Sun and identify the corona, chromosphere, photosphere, convection zone, radiation zone, and core.

☆ Describe the Sun's rotation and magnetic field.

☆ List the basic physical dimensions of the Sun.

☆ Describe some modern tools and techniques for studying the Sun.

☆ Describe the origin, properties, and cyclic nature of sunspots, and explain how sunspot variations are related to solar activity.

☆ Compare and contrast the origin and nature of solar granules, faculae, plages, flares, and prominences.

☆ Describe the origin and nature of the solar wind.

☆ Outline the puzzle of the missing solar neutrinos.

4.1 SUN AND EARTH

The Sun is the star closest to Earth. It provides the light, heat, and energy for life.

Ancient peoples worshipped the Sun as a life-giving god. Some of the

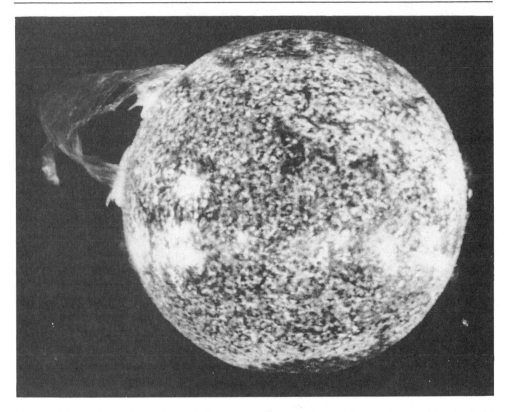

Figure 4.l. Solar activity. Several flares and a large arch prominence imaged in the light of ionized helium by a U.S. Skylab ultraviolet spectroheliograph. The big flare at the upper left spans more than 588,000 km across the Sun's surface.

names given to the Sun god were Aton, Apollo, Helios, and Sol. Scientists study the Sun today. It is critical to Earth and is a key to understanding distant stars that cannot be observed in detail.

The Sun's total energy output is enormous. The **Sun's luminosity L₀** is 3.85×10^{26} watts. Solar energy is practically inexhaustible. The amount of the Sun's energy that falls per second on Earth's outer atmosphere, called the **solar constant,** is about 1400 watts/m² (126 watts/square foot). This amount of energy provides about as much heat and light in a week as is available from all of our known reserves of oil, coal, and natural gas.

Our Sun is dynamic and seething (Figure 4.1). It is in turn extraordinarily active and relatively quiet. Changes in solar energy output affect Earth's climate, atmosphere, and weather, as well as modern power-transmission and communications systems. These changes are monitored to learn exactly how the Sun affects Earth.

State three reasons why modern astronomers, physicists, and engineers are using their most sophisticated techniques to determine the true nature of the Sun.

(1) _____

_____

(2) _____

_____

(3) _____

_____

Answer: (1) The Sun is an almost inexhaustible source of present and potential future energy. It is also free and nonpolluting! (2) The Sun is the only star close enough to observe in detail, so astronomers use it to determine what other stars are like. (3) Changes in the Sun's energy output affect Earth's climate, atmosphere, and weather, as well as power-transmission and communications systems.

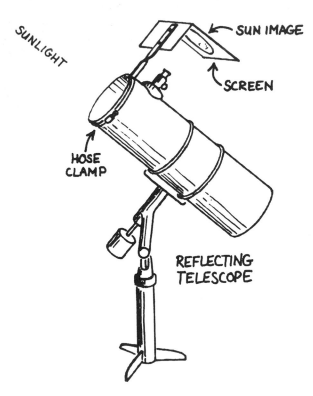

Figure 4.2. Projected image of the Sun focused on a screen behind the eyepiece of a small reflecting telescope. A way to avoid looking at the Sun while pointing a telescope is to use the shadow of the telescope cast on the screen as a guide.

4.2 DISTANCES AND SIZE

The average distance between Earth and Sun, called the **astronomical unit** (**AU**), is about 150 million km (93 million miles).

Astronomers calculate this distance from planetary data obtained by radar ranging. They use the astronomical unit as a measure of distance in the solar system (Table 8.1).

The Sun is a huge gaseous sphere. We see the apparent surface layer in the sky. Its **radius** (**R☉**) is about 696,000 km (432,000 miles). From Earth, the **Sun's angular diameter** = 32′ (about ½°) looks deceptively equal to the full Moon's. This illusion occurs because the Sun is 400 times farther away.

WARNING: *You could permanently blind your eyes if you observe the Sun without first taking proper precautions!* You should **never** look at the Sun directly and **never** look at the Sun through an optical instrument unless special solar filters properly cover the full aperture.

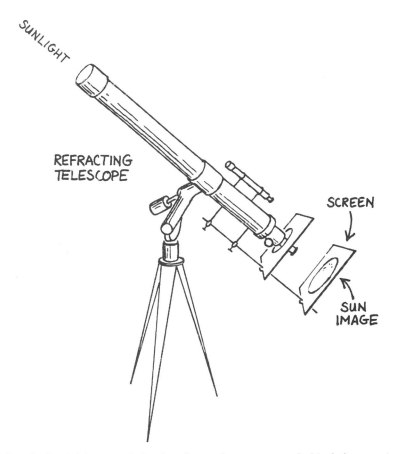

Figure 4.3. Projected image of the Sun focused on a screen behind the eyepiece of a small refracting telescope. A way to avoid looking at the Sun while pointing a telescope is to use the shadow of the telescope cast on the screen as a guide.

A way to observe the Sun is to project its image onto a screen and look only at the Sun's image on the screen (Figures 4.2 and 4.3).

About how many minutes does it take for sunlight to travel 1 AU? *Tip*: distance = speed × time. You can rewrite this: time = distance/speed. _____

Answer: About 8 .3 minutes. (That means that if the Sun stopped shining, you would not know about it until 8 .3 minutes later.)

Solution: The speed of light is ≈ 300,000 km (186,000 miles) per second and

$$1 \text{ AU} = 150,000,000 \text{ km}$$

$$\frac{150,000,000 \text{ km}}{300,000 \text{ km/second}} = \frac{93,000,000 \text{ miles}}{186,000 \text{ miles/second}}$$

$$= 500 \text{ seconds or } 8.3 \text{ minutes}$$

4.3 MAKEUP

The **nebular theory**, first proposed by German philosopher Immanuel Kant (1724–1804), says that our Sun and its planets formed together from a rotating cloud of interstellar gas and dust called the **solar nebula** about 5 billion years ago.

The solar nebula condensed into the newly forming Sun encircled by a rotating disk of gas and dust out of which the planets, moons, and other solar system objects formed (Figure 4.4). The Sun has more than 99 percent of the mass of the solar system and provides the gravitational force that keeps the planets circling it. Its surface gravity is practically 28 times Earth's.

More than 70 **chemical elements** have been identified in the Sun's spectrum. The Sun's outer layers likely have the same chemical composition as the Sun had at birth: about 73 percent hydrogen, 25 percent helium, and 2 percent other elements by weight. The Sun's core probably has subsequently changed to about 38 percent helium in nuclear fusion reactions.

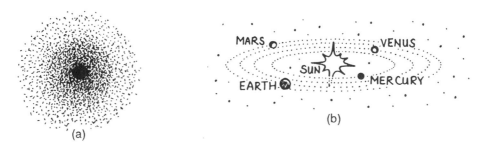

Figure 4.4.   Solar nebula theory. (a) A rotating nebula condensed to our Sun surrounded by a contracting disk where (b) the planetary system was born.

Why do astronomers expect to find other stars that have planets circling around them? _____

_____

_____

Answer: The nebular theory says that the planets circling the Sun were born together with their star. Since the Sun is a typical star, it seems likely that other, similar stars were also born together with a family of planets.

4.4 THE SUN'S STRUCTURE

Our picture of the Sun's structure comes from direct observations of its outer layers plus indirect theoretical calculations of the behavior of gases deep inside that we cannot see.

The three outer layers are called the Sun's **atmosphere**.

The **photosphere**, from the Greek "light ball," is the visible surface of the Sun. The photosphere is a hot, thin, opaque gas layer about 5800 K (10,000° F) from which energy is radiated into space. The **limb** is the apparent edge of the Sun's disk. It looks darker than the center, an effect called **limb darkening**, because light from the limb comes from higher, cooler regions of the photosphere.

The **chromosphere**, from the Greek "color ball," is a thin, transparent layer that extends about 10,000 km (6000 miles) above the photosphere. It is normally visible from Earth only during a total eclipse of the Sun, when it glows red due to its hydrogen gas. The temperature unexpectedly increases outward through the chromosphere, where the average temperature of matter is about 15,000 K.

The **corona**, from the Latin "crown," is the outermost atmosphere just above the chromosphere. It is a rarified, hot gas that extends many millions of kilometers into space. Because of its high temperature—up to 2 million K at the solar limb—the corona shines bright at X-ray wavelengths. During a total eclipse of the Sun, it is strikingly visible as a jagged white halo around the briefly hidden photosphere (Figure 4.5).

Below the photosphere is the Sun's **interior**. Theorists figure that temperature and density increase inward from the surface. No known element can survive as a solid or liquid at the extremely high solar temperatures. So the Sun must be made of very hot gases throughout.

Deep inside, the temperature must rise to 15 million K, the pressure to 200 billion atmospheres, and the density to over 100 times that of water. The core is the power plant where nuclear fusion reactions generate the Sun's energy (Section 5.5). There, hydrogen is fused into helium.

The intense energy released in the core provides heat inside the Sun and enough pressure to balance the inward pull of gravity. It is slowly transmitted outward. Photons are repeatedly absorbed and re-emitted at lower energies in the crowded **radiation zone**.

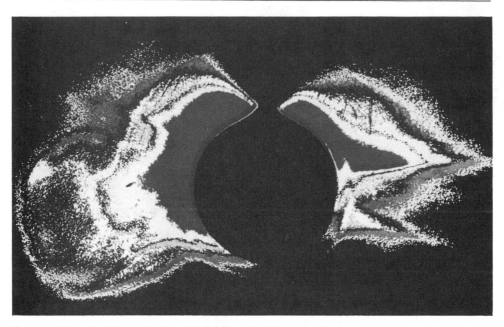

Figure 4.5. Coded to distinguish levels of brightness, the corona reaches outward for millions of kilometers in this image recorded by a U.S. Skylab coronagraph during a total eclipse of the Sun.

From there, circulating currents of gas in the **convection zone** transfer most of the energy as heat to the outer layers. It takes about 20 million years for energy produced in the core to surface and become sunshine.

Identify the regions of the Sun lettered on Figure 4.6. (a) _____ ; (b) _____ ; (c) _____ ; (d) _____ ; (e) _____ ; (f) _____

Answer: (a) Corona; (b) chromosphere; (c) photosphere; (d) convection zone; (e) radiation zone; (f) core.

4.5 ROTATION

The Sun keeps turning around its axis in space, from west to east, as Earth does. But there is a difference. All of Earth makes a complete turn in a day. The whole Sun does not turn around together at the same rate.

The **period of rotation**, or the length of time for one complete turn, is fastest at the Sun's equator (about 25 days), slower at middle latitudes, and slowest at the poles (about 35 days). This strange rotation pattern probably contributes to the violent activity that takes place on the Sun, described in the sections that follow.

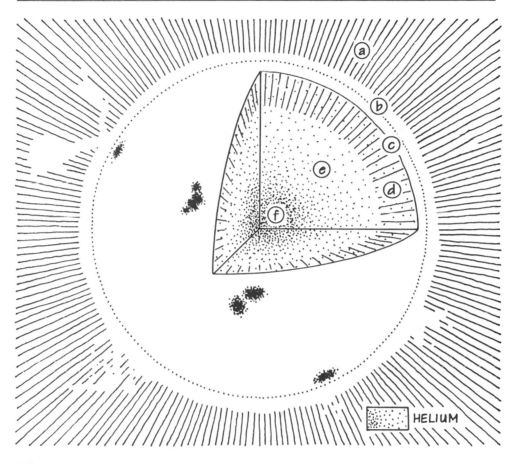

Figure 4.6.   Regions of the Sun.

How is it possible for different parts of the Sun to rotate at different rates, in contrast to Earth, all of which makes a complete turn in a day? _____

Answer: The Sun is a gaseous sphere and not a rigid solid as is Earth.

4.6 DATA

Summarize the data you have on the Sun's properties by filling in the convenient reference Table 4.1.

Answer: (a) About 150 million km (93 million miles); (b) 32′; (c) 1,390,000 km (864,000 miles); (d) 2×10^{30} kg; (e) 1.4 g/cm^3; (f) 1400 watts/m^2 (126 watts/ft^2); (g) 3.85×10^{26} watts; (h) about 5800 K; (i) G2; (j) –26.72; (k) 4.8; (l) equator: about 25 days; poles: about 35 days; (m) outer layers: about 73 percent hydrogen, 25 percent helium, 2 percent more than 70 other elements by weight; (n) 28 times Earth's or 294 m/s^2.

TABLE 4.1 Properties of the Sun

Quantity	Method of Measurement	Value
(a) Average distance from Earth	Radar ranging of planets	_____
(b) Angular diameter in sky	Solar telescope	_____
(c) Diameter	Angular diameter and distance	_____
(d) Mass	Planets' orbital motions	_____
(e) Average density	Mass and volume	_____
(f) Solar constant (solar energy incident on Earth)	High-altitude aircraft	_____
(g) Luminosity	Solar constant and distance from Earth	_____
(h) Surface temperature	Luminosity and radius	_____
(i) Spectral type	Spectrograph	_____
(j) Apparent magnitude	Photometer	_____
(k) Absolute magnitude	Apparent magnitude and distance from Earth	_____
(l) Rotation period	Sunspots' motions; Doppler shift	_____
(m) Chemical composition of outer layers	Sun's absorption spectrum	_____
(n) Surface gravity	Mass and radius	_____

4.7 OBSERVATIONS

Astronomers are using sophisticated tools and techniques to observe the Sun more closely and in more detail than ever before.

On the ground, special **optical solar telescopes** photograph the Sun's visible surface with its changing features (Figure 4.7). Arrays of giant **radio telescopes** receive and record radio waves from different parts of the radio Sun. **Infrared telescopes** observe the solar limb and map sunspots.

Color filters and **spectroheliographs** image the Sun in light of essentially a single wavelength. **Spectroheliograms** are images of the Sun in a single color light belonging to one gas such as hydrogen or calcium. They reveal the distribution of different gases and local phenomena (Figure 4.8).

In space, instruments monitor the Sun in all parts of the electromagnetic spectrum to detect solar features, radiations, particles, and fields normally blocked by Earth's atmosphere. **Ultraviolet, X-ray, and gamma ray telescopes** on spacecraft record images of processes in the hottest and most active regions of the Sun.

Formerly, the Sun's chromosphere and corona could be observed directly only during the few minutes of a total eclipse of the Sun when the much brighter photosphere was hidden. Now astronomers do not have to wait for one of these rare natural events to occur. **Coronagraphs**, telescopes designed to create an artificial eclipse, are used on the ground and in space to photograph the corona.

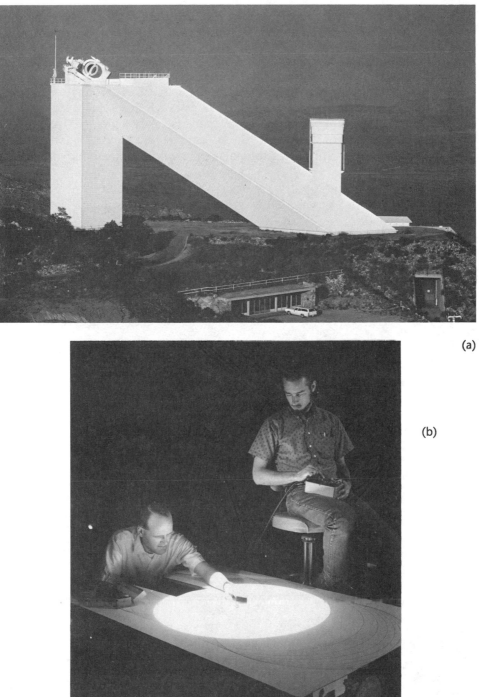

(a)

(b)

Figure 4.7. (a) The 1.5-m R.R. McMath-Pierce optical solar telescope at Kitt Peak. A mirror at the top reflects sunlight down the long sloping tube to (b) a room built inside the mountain where astronomers work with the Sun's image.

In 1973–1974, astronauts had a battery of eight solar telescopes aboard the orbiting U.S. Skylab space station 430 km (270 miles) above Earth. They observed the Sun extensively in visible and X-ray wavelengths. The Skylab coronagraph allowed 8 ½ months of corona observation compared to less than 80 hours from all natural eclipses since the use of photography began in 1839.

Violent solar eruptions were scrutinized in the International Solar Maximum year 1980–1981 project. The U.S. robot Solar Maximum Mission (SMM), the first satellite ever repaired in space by astronauts, completed historic optical, ultraviolet, X-ray, and gamma ray observations of solar flares in 1985. Experts require many years to analyze the voluminous data from space-based observations.

The European-U.S. robot spacecraft Ulysses is en route to observe the Sun's polar regions, magnetic fields, streams of particles and radiation, and

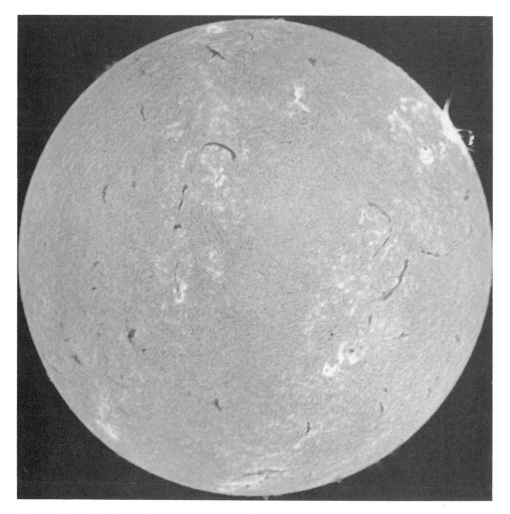

Figure 4.8. The Sun with a big east limb flare imaged in light of a spectral line of hydrogen, H_α, centered at 6563 Å.

environment in all possible solar latitudes in 1994–1995. Ulysses is the first spacecraft to have a flight path nearly perpendicular to the ecliptic.

Why do different features of the Sun appear in pictures taken in light of different wavelengths such as visible light, ultraviolet rays, or X-rays? *Tip*: Review Section 2.10 if necessary. _____

_____

Answer: Different wavelengths are produced in regions of different temperatures where different conditions and activities prevail.

4.8 SEETHING SURFACE

Optical telescopes reveal that the photosphere has a grainy appearance, called **granulation**. Bright spots that look like rice grains, called **granules**, dot the Sun's disk in high-resolution images (Figure 4.9).

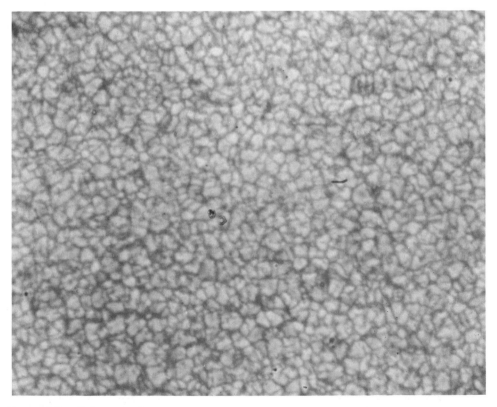

Figure 4.9. Solar granules.

Granules, cells up to 1000 km (625 miles) across, are the tops of rising currents of hot gases from the convection zone. Individual granules last an average of 5 minutes each. They look brighter than neighboring dark areas because they are about 300° hotter. The dark areas are descending currents of cooler gases.

Granules belong to **supergranules,** which are large, organized convection cells up to 30,000 km (19,000 miles) across, on the Sun's disk. Supergranules last several hours. They have a flow of gases from their centers to their edges, in addition to the vertical gas currents in the granules.

Spicules, jets of gas up 10,000 km (6000 miles) tall and 1000 km (600 miles) across, rise like fiery spikes into the chromosphere around the edges of supergranules. They change rapidly and last about 5 to 15 minutes.

Bright, white surface patches, called **faculae,** from the Latin "little torches," may be visible near the Sun's limb. Their appearance seems to signal coming solar activity.

What causes granulation? _____

Answer: Gases rising from the Sun's hot interior.

4.9 SUNSPOTS

Sunspots are temporary, dark, relatively cool blotches on the Sun's bright photosphere. They usually appear in groups of two or more. Individual sunspots last anywhere from a few hours to a few months.

The largest sunspots are visible at sunrise or sunset or through a haze. Observations of sunspots were first recorded in China before 800 B.C.

A typical sunspot is roughly as big as Earth. The largest sunspots may be bigger than ten Earths.

Sunspots really shine brighter than many cooler stars. They look dark only in comparison to the hotter, dazzling surrounding photosphere. The temperature is about 4200 K in the **umbra,** or core. The **penumbra,** or outer gray part of a large spot, is a few hundred degrees cooler than the photosphere.

Frequently sunspots appear in groups, or **solar active** regions, where the most violent solar activity occurs. The first telescopic observations of sunspots and their motions, reported by Galileo in 1610, had an historic impact (Section 8.7). Galileo correctly concluded that the Sun's rotation carries sunspots around.

Identify the umbra, penumbra, and photosphere, lettered on Figure 4.10, and indicate the approximate temperature of the umbra. (a) _____ ; (b) _____ ; (c) _____

Answer: (a) Photosphere; (b) penumbra; (c) umbra, 4200 K.

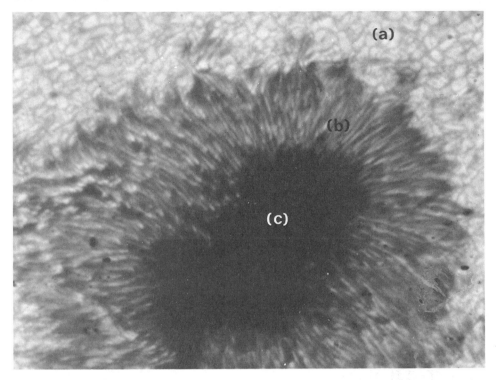

Figure 4.10. Umbra and penumbra of a sunspot, with granulation in the surrounding photosphere.

4.10 ACTIVITY CYCLES

At any one time more than 200 sunspots—or none at all—may appear on the Sun's disk. The number of sunspots regularly rises to a maximum and falls to a minimum in an approximately 11-year cycle, called the **sunspot cycle**.

The sunspot cycle, a display of the **solar activity cycle**, is watched carefully from Earth (Figure 4.11). The Sun is most active with greatest outbursts of energy and radiation for about 4 years, during which sunspots are most numerous, *sunspot maximum*. Record activity occurred last in 1990–1991 when sunspot numbers soared. The Sun is least active in the years of fewest sunspots, **sunspot minimum**.

Solar astronomers cannot predict coming solar activity. They aim to understand the solar activity cycle in order to make reliable forecasts.

Why is it important to keep track of the sunspot cycle? _____

_____

_____

Answer: The Sun is most active during the years of sunspot maximums, pouring the greatest amount of energy and radiation into Earth's environment.

MINIMUM   MAXIMUM

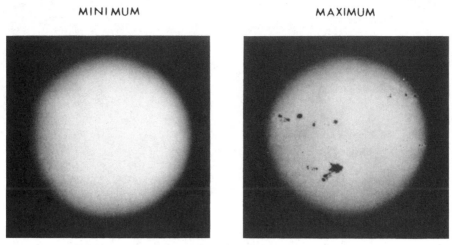

16 June 86   13 June 89

Figure 4.11. White-light photograph of the Sun at two different times in its activity cycle.

4.11 MAGNETISM

Sunspots are like huge magnets. These regions of powerful magnetic fields are typically thousands of times stronger than Earth's magnetic field.

The magnetic field of a sunspot can be detected before the spot itself can be seen and after the spot is gone. Therefore, magnetic fields probably shape and control local conditions on the Sun. Astronomers analyze magnetic fields by measuring Zeeman spectral line-splitting (Section 3.10).

A weaker magnetic field spreads out over the whole Sun. It has a north magnetic pole and a south magnetic pole, with the magnetic axis tilted 15° to the rotation axis. It is split into two hemispheres. A display of magnetic field strength is called a **magnetograph**.

The Sun's magnetic field probably extends from its northern hemisphere through the solar system out to Pluto about 6 billion km (4 billion miles). Near the edge of the solar system, the magnetic field bends and returns to the Sun's southern hemisphere.

The complex solar magnetic field is generated by rotational and convective motions of electrically charged particles that make up the Sun's hot gases. Apparently it energizes and controls violent outbursts of material and radiation on the Sun.

The polarity of the Sun's magnetic field is reversed about every 11 years shortly after the period of sunspot maximum. It takes two sunspot cycles of about 11 years each for the Sun's magnetic poles and sunspot magnetic polarities to repeat themselves. So the solar activity cycle is 22 years when counting the length of time required for the Sun to return to its original configuration.

What probably activates the violent outbursts of material that occur on the Sun? _____

Answer: Very strong magnetic fields at the sites of sunspots.

You can observe a magnetic field by putting a magnet under a piece of paper. Lightly sprinkle iron filings on top of the paper. The filings will line up according to the strength of the magnetic force. By showing the regions of the magnetic force, they make the magnetic field visible to you.

What probably bends and controls the trajectory of the ejected gas in solar flares? _____

Answer: Strong magnetic fields in the vicinity of sunspots.

4.12 FLARES AND PROMINENCES

A solar **flare** is a sudden, tremendous, explosive outburst of light, invisible radiation, and material from the Sun. One great solar flare may release as much energy as the whole world uses in 100,000 years (Figure 4.12).

Flares are short-lived, typically lasting a few minutes. The largest last a few hours. They occur near sunspots, especially in periods of sunspot maximums. Flares seem to be energized by strong local magnetic fields (Figure 4.13).

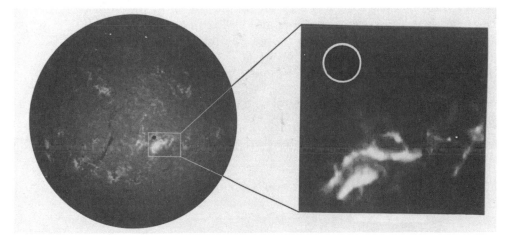

Figure 4.12. Solar flare nearly 300,000 km (180,000 miles) across and covering over 5.2 billion square km (2 billion square miles) of the Sun's surface, from the U.S. robot Solar Maximum Mission satellite. Inset at right, magnified 20 times, shows ultraviolet emission from the flare. The black dot represents Earth for size comparison.

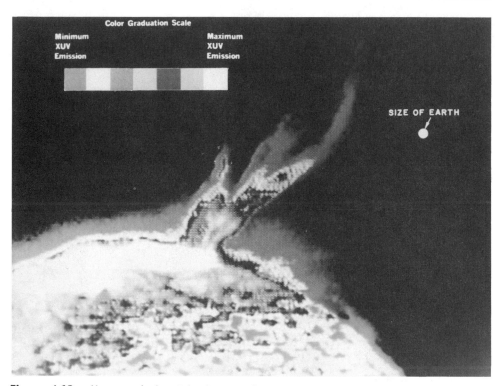

Figure 4.13. X-ray and ultraviolet image of a solar flare, coded to distinguish levels of brightness. The white dot represents Earth for size comparison.

Prominences are fiery arches of ionized gases on the limb of the Sun. They can last for weeks or months and rise tens of thousands of kilometers up. Apparently prominences are supported and twisted by magnetic fields in the vicinity of sunspots (Figure 4.1).

What probably bends and controls the trajectory of the ejected gases in solar flares and prominences? _____

_____

Answer: Strong magnetic fields in the vicinity of sunspots.

4.13 HOW SOLAR FLARES AFFECT EARTH

A huge flare can hurl fantastic amounts of high-energy radiation and electrically charged particles—as much energy as a billion exploding hydrogen bombs—into the solar system.

Flare gamma rays, X-rays, and ultraviolet rays reach Earth in just 8 minutes. Flare particles arrive a few hours or days later. These could destroy all life on Earth if our planet were not shielded by its magnetic field and atmosphere. Travelers in supersonic airplanes and spacecraft must be protected, too.

When high-energy particles from the Sun strike Earth's atmosphere, they can stimulate the atmospheric atoms and ions to radiate light, producing auroras.

The **aurora borealis**, or northern lights, and **aurora australis**, or southern lights, are spectacular bands of light that sometimes shine in the night sky, mainly in Earth's Arctic and Antarctic regions, but occasionally also at midlatitudes. Maximum auroral activity occurs around Earth's magnetic poles. Auroras are visible about 2 days after a solar flare. They reach a peak about 2 years after a sunspot maximum.

Strong blasts of flare particles that interact with Earth's magnetic field can cause magnetic storms in which compasses don't work normally. Flares can also cause atmospheric storms, surges in power and telephone lines and blackouts.

Flare high-energy radiation heats the upper atmosphere, making it expand. Then friction and the drag on satellites in low orbits increase. The drag is greatest during times of maximum solar activity when satellites may plunge from orbit and be destroyed on re-entry. The U.S. Skylab in 1979 and Solar Maximum Mission in 1989 were casualties of solar maximums. By increasing ionization, flares can disrupt radio transmission.

Because solar flares can affect modern life directly, solar astronomers carefully monitor the Sun's magnetic field and activity daily. So far, no one can foretell exactly when a flare will occur. (Figure 4.14). High priority is placed on timely warnings of flares that will affect Earth.

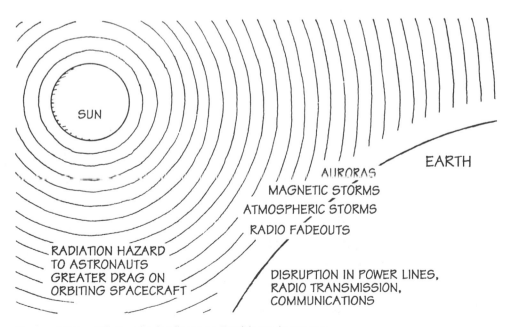

Figure 4.14. Effects of solar flares on Earth's environment.

List two effects that large solar flares have on modern technology on Earth.

(1) _____ ; (2) _____

Answer: (1) Disruption in power transmission; (2) disruption in radio communications.

4.14 SOLAR WIND

The **solar wind** is a plasma, or stream of energetic, electrically charged particles that flows out from the Sun at all times. It is much faster, thinner, and hotter than any wind on Earth.

The solar wind is observed by instruments carried on spacecraft above Earth's atmosphere. Near Earth, the average solar wind speed is about 450 km/second (1 million miles/hour). Travel time from Sun to Earth is about 4 days. Earth's atmosphere and magnetic field ordinarily protect us from harmful effects of the solar wind.

Big blasts of solar wind occur during solar flares. The wind is strongest during periods when many sunspots are visible and solar activity is great. Strong blasts of solar wind can produce especially brilliant auroras.

The solar wind comes mainly from **coronal holes**, regions in the Sun's corona where gases are much less dense than elsewhere. Magnetic fields are relatively weak there, allowing high-speed solar wind streams to escape.

Instruments on the Voyager spacecraft (Section 8.12) continue to measure the solar wind beyond the orbit of Neptune. They could detect the **heliopause**, the boundary where the solar wind effectively no longer exists.

What is the solar wind? _____

_____

Answer: A stream of energetic, electrically charged particles that flows out from the Sun.

4.15 PROBING THE INTERIOR

Until recently, scientists were sure they understood what makes the Sun shine. **Solar neutrino experiments** have raised some doubts.

Theoretically, the Sun's energy is produced by the conversion of hydrogen into helium in nuclear fusion reactions. **Solar neutrinos**, elementary particles whose main characteristic is that they interact very weakly with matter and pass freely through it, are also produced in these reactions.

Scientists cannot look directly deep inside the Sun's core to test their theory. But they predict that the neutrinos produced in the core should escape. So they look for the solar neutrinos instead.

If neutrinos were to be detected in the amount predicted by theory, they would provide evidence that the theory is correct.

Scientists have built neutrino traps deep inside the Earth. The number of neutrinos detected over the last 20 years in underground laboratories in the U.S., Europe, Japan, and Russia is lower than the number theory predicts. More independent, sensitive experiments and further analysis may explain this **solar neutrino problem.**

Helioseismology is the new study of the Sun's internal structure and condition by measuring global oscillations on its surface. Pressure waves on the Sun can reveal the density, temperature, and rotation rate inside the Sun just as earthquake waves expose Earth's interior. The solar oscillations are observed spectroscopically through Doppler shifts in certain spectral lines (Section 3.9). **Astroseismology** extends this study to other stars.

Give two possible explanations for the unexpectedly small number of solar neutrinos detected in experiments to date.

(1) _____ ;

(2) _____

Answer: Our understanding of (1) processes in the Sun's interior or (2) of neutrinos is incorrect or incomplete. (Astronomers rely on the results of the experiments.)

4.16 COMMON FEATURES

Apparently other stars have regions of violent activity like those of our Sun, including **starspots** and **starspot cycles**, although stars are so far away that these must be deduced from their spectral lines and brightness variations rather than observed directly. Recent X-ray observations indicate that nearly all types of stars also have similar coronas with temperatures of at least a million degrees.

Write a short summary describing three phenomena that indicate violent activity on the Sun (and by extension to other stars), and name their probable cause.

_____

_____

_____

Answer: Your answer should briefly describe (1) sunspots, or dark, relatively cool, temporary spots on the Sun's photosphere; (2) flares, or sudden, short-lived outbursts of light and material near a sunspot; (3) prominences, or fiery arches of ionized gases on the limb of the Sun.

Most violent activity on the Sun seems to be caused and controlled by very strong local magnetic fields.

4.17 MOTIONS IN SPACE

The Sun, like all other stars, is racing through space.

With respect to nearby stars, the Sun is speeding toward the constellation Hercules at 20 km/second (45,000 miles/hour), carrying its nine planets along with it.

The Sun with its planets is inside the Milky Way Galaxy. It goes around our Galaxy's center as the whole Galaxy turns around in space. The sun travels at about 250 km/second (563,000 miles/hour) (Section 6.2).

SELF-TEST

This self-test is designed to show you whether or not you have mastered the material in Chapter 4. Answer each question to the best of your ability. Correct answers and review instructions are given at the end of the test.

1. List three reasons why modern astronomers study the Sun.

   (1)_____

   _____

   _____

   (2)_____

   _____

   _____

   (3)_____

   _____

   _____

2. Match the most appropriate tool to the work:

   ____ (a) Image processes in the hottest active regions of the Sun.

   ____ (b) Photograph corona outside solar eclipse.

   ____ (c) Photograph the Sun's visible surface.

   ____ (d) Photograph the Sun in the light of a particular element.

   ____ (e) Receive and record solar radio waves.

   (1) Coronagraph.
   (2) Optical solar telescope.
   (3) Radio telescopes.
   (4) Spectroheliograph.
   (5) Ultraviolet ray, X-ray, and gamma ray telescopes.

3. Define the astronomical unit (AU). _____

   _____

4. Sketch the Sun, and identify the corona, chromosphere, photosphere, convection zone, radiation zone, and core.

5. Estimate (a) diameter; (b) mass; and (c) surface temperature of the Sun.

   (a) _____ ; (b) _____ ; (c) _____

6. Why is the sunspot cycle carefully monitored from Earth?_____

_____

_____

_____

_____

7. Identify the following phenomena of the Sun:

_____ (a) Low-density region in the corona where the solar wind originates.

_____ (b) Bright cell that looks like rice grain in the photosphere.

_____ (c) Dark, relatively cool blotches in the bright photosphere.

_____ (d) Elementary particles predicted to be produced in nuclear reactions in the core.

_____ (e) Tremendous, short-lived, explosive outburst of light and material.

(1) Flare.
(2) Granule.
(3) Coronal hole.
(4) Solar neutrino.
(5) Sunspot.

8. What is the solar wind?_____

_____

9. List four ways that a flare and unusually big blasts of solar wind can affect Earth's environment. (1)_____

(2)_____

(3)_____

(4)_____

10. (a) What is the solar constant? (b) Why is it important to know if it is truly constant or if it varies with time? _____

_____

_____

_____

_____

ANSWERS

Compare your answers to the questions on the self-test with the answers given below. If all of your answers are correct, you are ready to go on to the next chapter. If you missed any questions, review the sections indicated in parentheses following the answer. If you missed several questions, you should probably reread the entire chapter carefully.

1. (1) The Sun is a free, nonpolluting, almost inexhaustible source of present and potential future energy.

   (2) The Sun is the only star close enough to observe in detail, so astronomers use it to determine what other stars are like.

   (3) Changes in the Sun's energy output affect Earth's climate, atmosphere, and weather, as well as power-transmission and communications systems. (Sections 4.1, 4.13)

2. (a) 5;   (b) 1;   (c) 2;   (d) 4;   (e) 3.   (Section 4.7)

3. The astronomical unit (AU) is the average distance between the Earth and the Sun, about 150 million km (93 million miles) (officially 149,597,870 km). (Section 4.2)

4. See Figure 4.6, Regions of the Sun. (a) Corona;   (b) chromosphere;   (c) photosphere;   (d) convection zone;   (e) radiation zone;   (f) core. (Section 4.4)

5. (a) 1,390,000 km (864,000 miles);   (b) 2×10^{30} kg;   (c) 5800 K (10,000° F).   (Sections 4.2, 4.4, 4.6)

6. The sunspot cycle is watched carefully from Earth as an indicator of solar activity. The Sun is most active, with greatest outbursts of energy and radiation, during the years when sunspots are most numerous. It is least active in the years of sunspot minimums.   (Sections 4.10, 4.13, 4.14)

7. (a) 3;   (b) 2;   (c) 5;   (d) 4;   (e) 1.   (Sections 4.8, 4.9, 4.12, 4.14, 4.15)

8. A stream of energetic electrically charged particles that flows out from the Sun.   (Section 4.14)

9. (1) Increased hazardous radiation;   (2) auroras;   (3) magnetic storms; (4) atmospheric storms.   (Sections 4.13, 4.14)

10. (a) The amount of the Sun's energy that falls per second on Earth's outer atmosphere, about 1400 watts per square meter (126 watts per square foot). (b) Changes in the solar constant might drastically change Earth's climate and atmosphere.   (Sections 4.1, 4.13)

5

STELLAR EVOLUTION

*To every thing there is a season, and a
time to every purpose under the heaven.
A time to be born, and a time to die*

Ecclesiastes 3:1–2

Objectives

☆ Define stellar evolution.

☆ List the stages in the life cycle of a star like our Sun according to the modern theory of stellar evolution.

☆ Explain the importance of the H–R diagram to theories of stellar evolution.

☆ Explain the relation between a star's age and its position on the H–R diagram.

☆ List the three main steps in the birth of a star.

☆ Describe the energy balance and pressure balance in main sequence stars.

☆ Compare and contrast what happens in the advanced stages of evolution for stars of large and small mass: planetary nebulas, white dwarfs, supernovas, pulsars/neutron stars, and black holes.

☆ Identify nebulas, main sequence, blue giant, red giant, and pulsating variable stars that can be observed in the sky.

☆ Explain how supernovas and pulsars are observed.

☆ Describe the origins of the different chemical elements and the importance of supernovas to new generations of stars.

☆ Describe observational evidence for black holes.

5.1 LIFE CYCLE OF STARS

No star shines forever. **Stellar evolution** refers to the changes that take place in stars as they age—the life cycle of stars. These changes cannot be observed directly, because they take place over millions or billions of years. Astronomers construct a theory of stellar evolution that is consistent with the laws of physics. Then they check their theory by observing real stars shining in the sky.

In checking theory against observations, astronomers make use of H–R diagrams. Theoretical predictions are made regarding a sequence of changes in luminosity and temperature for stars as they go from birth to death. These changes are plotted on an H–R diagram, forming theoretical **tracks of evolution**. Theoretical H–R diagrams are then compared with H–R diagrams for groups of real stars (Section 6.4).

The predictions of the **modern theory of stellar evolution**, described in this chapter, agree well with the data from observations of real stars.

What is stellar evolution? _____

_____

Answer: The changes that take place in stars as they age—the life cycle of stars.

5.2 BIRTHPLACES

Stars form out of matter that exists in space. The gigantic **interstellar** (between stars) **clouds** of gas and dust must be the birthplaces of stars.

You can see the nearest cloud in space where new stars are forming now. The famous Orion Nebula, located about 1500 light-years away in the constellation Orion, is a region of intense star formation (Figure 5.1).

Look for the Orion Nebula in the winter. It is marked on your winter skies

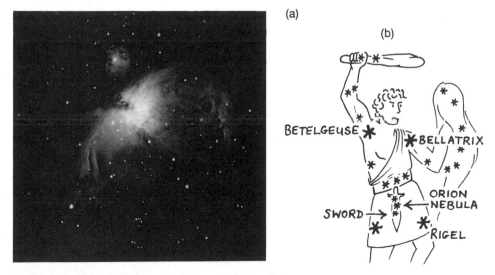

(a)

(b)

BETELGEUSE
BELLATRIX
ORION NEBULA
SWORD
RIGEL

Figure 5.1.   Orion Nebula, in the constellation Orion.

map in the sword of Orion the Hunter. The Orion Nebula looks like a hazy patch to your eye. Through a telescope you will see it glow with a greenish color. Hot, newly formed stars in the region make the gases glow. A much larger associated cloud is not visible.

Are new stars still being born today? Where? _____

_____

Answer: Yes. In gigantic clouds of gas and dust, as in the Orion Nebula.

5.3 BIRTH

A **protostar** is a star in its earliest phase of evolution. You can think of a protostar as a star that is being born.

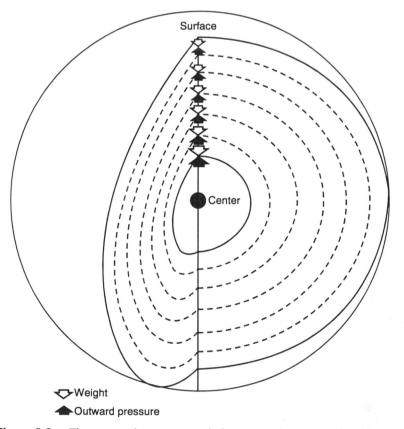

Figure 5.2. The outward gas pressure balances gravity at every level in a star.

Protostars form by chance at high-density clumps inside huge turbulent gas (mostly hydrogen) and dust clouds that exist in space. Perhaps a shock wave from an exploding star (supernova) triggers the process.

A protostar is held together by the force of gravity. Initially, the force of gravity pulls matter in toward the center of a dense clump, causing it to contract and become even denser. Matter continues to accrete onto the protostar as it contracts. Gravitational contraction of the cloud and protostar causes the temperature and pressure inside to rise greatly.

Heat flows from the protostar's hot center to its cooler surface. The protostar radiates this energy into space. It shines at infrared wavelengths.

In a rotating cloud, a disk of dust and gas may surround a protostar. This disk also reradiates the energy as infrared. Possibly particles in the disk accrete to form planets (Figure 12.2).

When the temperature in the protostar's center reaches 10 million K, nuclear fusion reactions start. These nuclear reactions release tremendous amounts of energy. Energy is generated in the center as fast as it is being radiated out into space. The very high internal temperatures and pressures are thus maintained.

The outward pressure of the very hot gases balances the inward pull of gravity (Figure 5.2). This balance is called **hydrostatic equilibrium**. The protostar stops contracting. It shines its own light steadily into space. The protostar becomes a newborn star. Most likely our Sun was born in this way about 5 billion years ago.

Recent observations support this theory of star birth. Protostars in dense cores of gaseous clouds are imaged at infrared wavelengths. Jets of gas are seen streaming away from young stars. They may be aligned by a planet-forming circumstellar disk.

List the three main steps in the birth of a star. (1) _____

_____ ;

(2) _____ ;

(3) _____

Answer: (1) Gravitational contraction within a cloud of gas and dust; (2) rise in interior temperature and pressure; (3) nuclear fusion.

5.4 LIFETIMES

The clouds in which protostars form do not have identical masses or distributions of the chemical elements. The life cycle of a star—the time it takes for a star to evolve—depends upon its initial **mass** and **chemical composition**.

Stars that begin life with about the same mass and chemistry go through the same stages of evolution in about the same amount of time.

Stars of similar chemistry with very high mass evolve fastest, while those of very low mass take the longest time to evolve.

The theoretical evolutionary tracks on the H–R diagram in Figure 5.3 show how a protostar's luminosity and temperature change as it contracts to become a star.

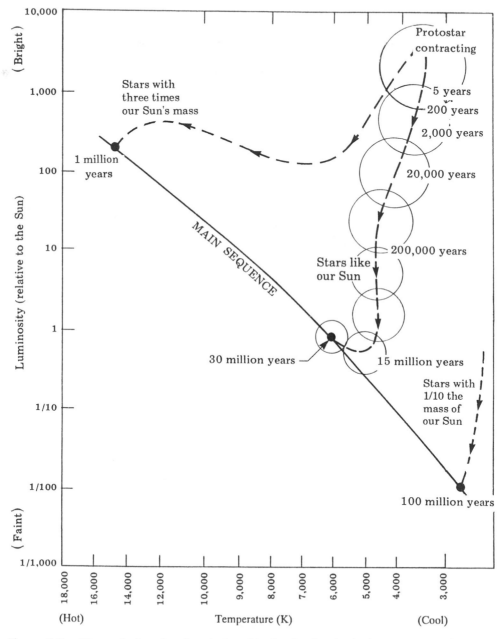

Figure 5.3. Theoretical tracks of evolution showing luminosity and temperature changes of contracting protostars of different masses. (Contraction times are marked at the tracks' endpoints.)

Approximately how long does it take each of the following protostars to reach zero age main sequence (to be born)? (a) stars like our Sun _____ _____ ; (b) stars with mass much greater than the Sun's _____ ; (c) stars with mass much less than the Sun's _____

Answer: (a) About 30 million years; (b) about 1 million years; (c) about 100 million years.

5.5 WHY STARS SHINE

You can think of a **main sequence** star as an adult star. In comparison to changes in protostars, evolution of main sequence stars is very slow. A star spends most of its lifetime shining steadily, with luminosity and temperature values found along the main sequence of H–R diagrams.

A main sequence star gets its energy from **nuclear fusion reactions** in which hydrogen at the center of the star is converted into helium (Figure 5.4). Four hydrogen nuclei are fused into one lighter, helium nucleus. The disappearing mass is changed into energy and released. (The same process releases energy in hydrogen bombs.)

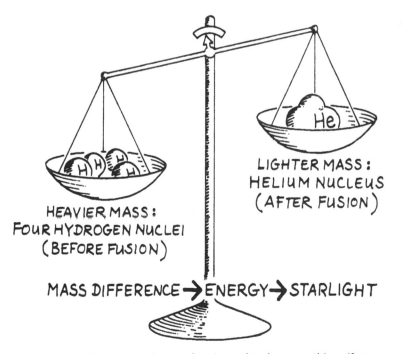

Figure 5.4. An imaginary experiment showing why the stars shine. If you could weigh the hydrogen nuclei before and the helium nucleus after fusion, you would discover that the helium nucleus was lighter.

The energy from the nuclear fusion reactions eventually reaches the star's surface. Then the star shines energy into space.

The amount of energy released in a nuclear fusion reaction can be calculated from the famous equation of (German-born) U.S. physicist Albert Einstein:

$$E = mc^2$$

where E = energy, m = mass difference, and c = speed of light.

According to Einstein's equation, when many nuclear fusion reactions occur together, enormous amounts of energy are released. The Sun is a huge, hot gaseous sphere that shines steadily without appreciable change of size or temperature. Although practically 5 million tons of hydrogen must be converted into helium each second to produce the Sun's luminosity, less than 0.01 percent of the Sun's total mass changes to sunshine in a billion years.

What is the source of energy that lets main sequence stars shine?

_____

Answer: Nuclear fusion reactions in which hydrogen is converted into helium.

5.6 OLD AGE ☆

A star will shine steadily as a main sequence star until all the available hydrogen in its core has been converted into helium. Then the star will begin to die.

Our Sun is an average medium-sized star. It has been shining as a stable main sequence star for about 5 billion years, and it should continue to shine steadily for another 5 billion years.

Very massive, hot, bright stars die fastest because they use up their hydrogen most rapidly. The very massive blue giant stars, such as Rigel in Orion, spend only a few million years shining as main sequence stars.

The least massive, cool, dim stars live the longest because they consume their hydrogen fuel least rapidly. The small-mass red dwarfs are the oldest and most numerous main sequence stars. They have lifetimes billions of years long.

What types of stars are expected to live (a) longest? _____

_____(b) shortest? _____ (c) About how

much longer is the Sun expected to shine as it does now? _____

_____

Answer: (a) Those with small mass, such as red dwarfs; (b) very massive stars, such as blue giants; (c) about 5 billion years.

5.7 RED GIANTS

After the hydrogen fuel in the star's core is used up, the star no longer has an energy source there. The core, which then consists primarily of helium, begins to contract gravitationally. Hydrogen fusion continues at the boundary between the helium core and the outside envelope of hydrogen.

Gravitational contraction causes the temperature of the helium core to rise. The high temperature makes the boundary hydrogen fuse faster, and the star's luminosity increases.

The tremendous energy released by this hydrogen fusion and gravitational contraction heats up surrounding layers. The star expands to gigantic proportions. The star's density is then very low everywhere except in the core (Figure 5.5).

As the star expands, its surface temperature drops and its surface color turns to red. The star has changed into a huge, bright, red, aging star—a **red giant**. It is cool but bright because of its gigantic surface area. It has the luminosity and temperature values of the red giant region of the H–R diagram.

You can see some red supergiant stars shining in the sky. Good examples are Betelgeuse in Orion and Antares in Scorpius, both over 400 times the Sun's diameter (Tables 1.1 and 2.1).

Our Sun, like all stars, is expected to change into a huge red giant when it dies. That red giant Sun will shine so brightly that rocks will melt, oceans will evaporate, and life as we know it on Earth will end.

When does a star begin to change from a main sequence star into a red giant?

Answer: When it has converted all of the available hydrogen fuel in its core into helium.

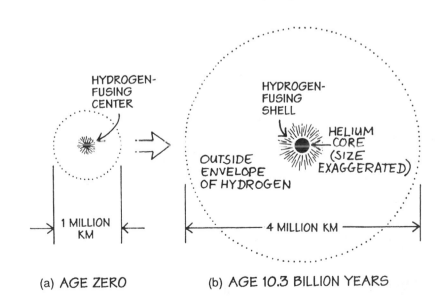

(a) AGE ZERO   (b) AGE 10.3 BILLION YEARS

Figure 5.5.   Sunlike star (a) at start of life on main sequence and (b) as it ages to red giant.

5.8 SYNTHESIS OF HEAVIER ELEMENTS

Gravitational contraction causes the temperature inside the red giant's helium core to rise to 100 million K. At that temperature, helium is converted to carbon in nuclear fusion reactions (Figure 5.6).

The helium core does not expand much once the helium fusion starts. The temperature builds up rapidly without a cooling, stabilizing expansion. Helium nuclei fuse faster and faster, and the core gets even hotter. This nearly explosive ignition of helium fusion is called the **helium flash**.

After some time, the temperature rises sufficiently so that the core expands. Cooling occurs inside, and helium fusion goes on at a steady rate, surrounded by a hydrogen-fusing shell.

Inside the more massive red giants, further fusion reactions can build up familiar elements heavier than carbon, such as oxygen, aluminum, and calcium (Appendix 4).

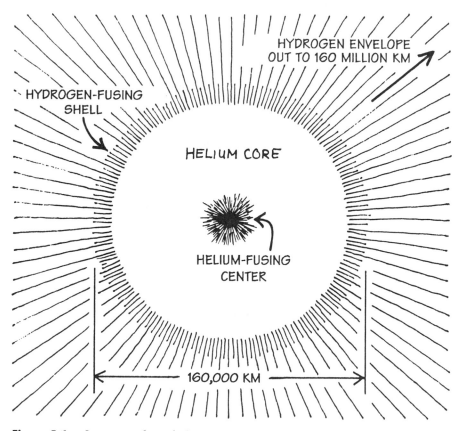

Figure 5.6.   Structure of a red giant star.

Astronomers believe that elements like carbon and oxygen, which we need for life, are made where? _____

Answer: Inside red giant stars.

5.9 VARIABLE STARS

A star probably moves back and forth between the red giant region and the main sequence several times, in a way not yet fully understood, before it enters the final stages of its life.

Most stars probably change from red giants to **pulsating variable stars** before they finally die. That is, they expand and contract and grow bright and fade periodically.

Cepheid variables are very large luminous yellow stars whose light output varies in periods of from 1 to 70 days. You can observe Delta Cephei, the first discovered and the star for which this class of variables was named (Figure 5.7). Cepheids are important because they provide a way of measuring distances too great to be measured by trigonometric parallax.

More than 700 Cepheid variables are known in the Milky Way Galaxy. Polaris, the North Star, is the nearest. Its brightness varies between magnitudes 2.5 and 2.6 about every 4 days.

U.S. astronomer Henrietta Leavitt (1868–1921) discovered that the longer

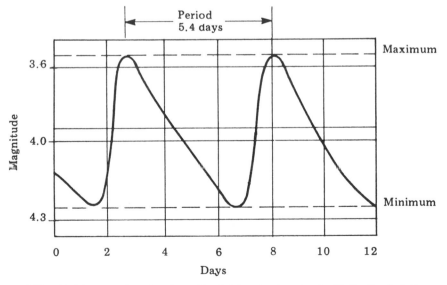

Figure 5.7.   Light curve showing how the light output varies for Delta Cephei, the prototype Cepheid variable star.

the period of light variation of Cepheids, the greater the luminosity. Astronomers use this **period-luminosity relation** to determine the absolute magnitude of Cepheids after measuring their periods.

A comparison of the calculated absolute magnitude and the observed apparent magnitude yields the distance to the Cepheids and the star groups they belong to (Section 3.16). Cepheids are useful **distance markers** out to about 3 Mpc (10 million light-years).

RR Lyrae variables, named after variable star RR in the constellation Lyra, are pulsating blue-white giants whose light output varies from brightest to dimmest in periods of less than a day. About 4500 RR Lyrae stars are known in the Milky Way Galaxy. RR Lyrae stars are used to measure the distance to the star clusters they belong to, out to about 200,000 pc (600,000 light-years).

Long-period Mira variables, named for famous Mira in the constellation Cetus, are red giants that take between 80 and 1000 days to vary between brightest and faintest. Mira, about 40 pc (130 light-years) away, varies from its maximum bright red to its minimum output, where it becomes invisible, in a period of 332 days. Mira was named the "Wonderful" by amazed seventeenth-century observers, who first recorded its brightness fluctuations.

What two characteristics of a pulsating variable star change periodically?
(1) _____ ; (2) _____

Answer: (1) Size; (2) luminosity.

5.10 DEATH

All stars evolve in about the same way, although over different periods of time, until their cores become mostly accumulated carbon (Figure 5.8). The last stage in a star's evolution, or the way it finally dies, depends greatly on its mass.

Small stars, up to about 1.4 times the Sun's mass, finally die without a fuss, quietly fading away into the blackness of space. Very massive stars end with a violent explosion, flaring up brilliantly before giving up life.

What characteristic of a star determines the way it finally dies? _____

Answer: Its mass.

5.11 MASS LOSS

When a star of mass like our Sun has depleted all of its available helium fuel, it becomes a bloated red giant star for the last time. (At this stage of its life our Sun will become so big that it will swallow up Mercury, Venus, Earth, and Mars.)

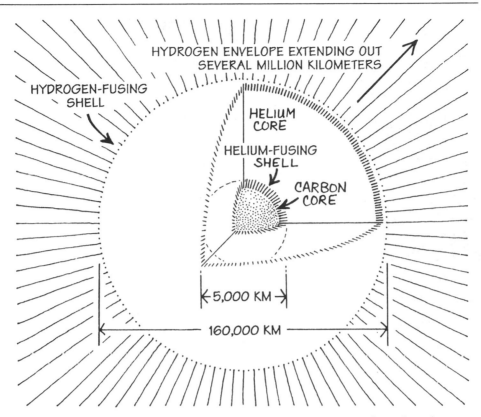

Figure 5.8. The structure of a star with an increasing inner core of mostly carbon.

The star then throws off some of its mass. The star's outermost hydrogen envelope, enriched by heavier elements, flies off into space. Electrically charged particles stream away in a flow called a **stellar wind**. (The solar wind is described in Section 4.14). Deeper layers are thrown off in a wispy, expanding shell of gas typically about 0.5 to 1 light-year across, called a **planetary nebula**, which continues to spread out at speeds of about 20 to 30 km/sec (45,000 to 67,500 miles/hour). The star's core is left behind.

About 1600 planetary nebulas have been recorded. They are probably less than 50,000 years old, because the gas atoms in the nebula separate rapidly. After about 100,000 years, the shell is too spread out to be visible.

Examine Figure 5.9. Identify the core of the star and the planetary nebula in the photograph. (a) _____ ;

(b) _____

Answer: (a) Planetary nebula; (b) core of the star.

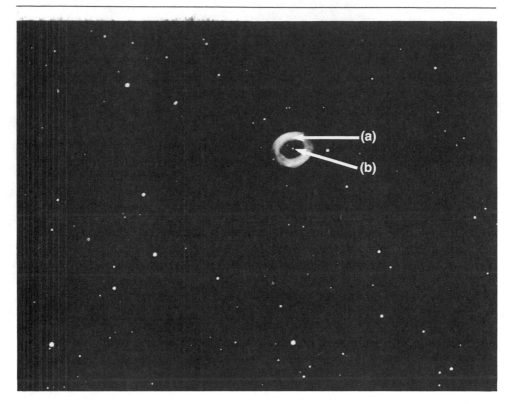

Figure 5.9.   Famous Ring Nebula in Lyra, a planetary nebula and star core.

5.12 WHITE DWARFS

After it has thrown off its gas envelope, the star remains as a core of carbon surrounded by a shell of burning helium.

A star that has exhausted all of its nuclear fuel can no longer withstand the pull of gravity. It contracts again as gravity pulls matter in toward the center. Gravitational contraction makes the temperature and pressure go up very high, and electrons are stripped off atoms. The star becomes a small, hot, **white dwarf**. It is made mostly of electrons and nuclei. These subatomic particles can be squeezed much closer together than whole atoms can.

Eventually, when the white dwarf star reaches about Earth's size, it cannot contract any further. White dwarf stars of mass like the Sun are very dense because gravity packs all that mass into a star the size of Earth. The force of gravity on such a white dwarf star would be about 350,000 times greater than that on Earth. If you could stand on a white dwarf star, you would weigh 350,000 times more than you do on Earth.

Sometimes at this stage a **nova**, a brilliant flaring star, is produced. If the white dwarf belongs to a binary system, matter from its companion star may fall onto the white dwarf and fuel the brief, bright flare.

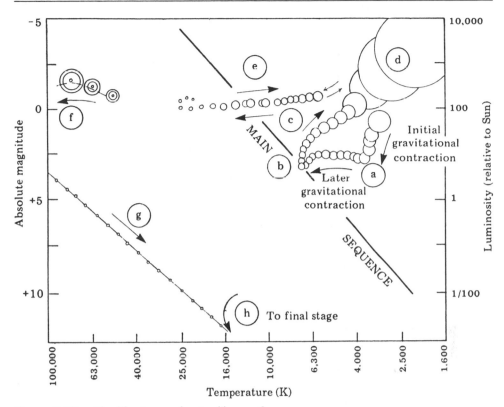

Figure 5.10. The life stages of a star like our Sun.

Gradually the white dwarf star cools, turns to dull red, and shines its last energy into space. Then the white dwarf becomes a dead **black dwarf** in the graveyard of space.

What is a white dwarf star? _____

_____

_____

Answer: A small dense (dying) star of low luminosity and high surface temperature, typically about the size of Earth but with mass equal to the Sun's.

5.13 LIFE CYCLE OF SUNLIKE STARS

Identify each stage of the life of a star like our Sun, as labeled sequentially in Figure 5.10. (a) _____ ;

(b) _____ ;

(c) _____ ;

(d) _____ ;

(e) _____ ;

(f) _____ ;

(g) _____ ;

(h) _____

Answer: (a) Protostar, gravitational contraction of cloud of gas and dust; (b) stable main sequence star, shining by nuclear fusion (converting hydrogen to helium); (c) evolution to red giant when helium core forms; (d) red giant, shining by helium fusion; (e) variable star, formation of carbon core; (f) planetary nebula, enriched hydrogen envelope ejected into space; (g) white dwarf, mass packed into star about the size of Earth; (h) dead black dwarf in space.

5.14 EXPLODING STARS

Very massive stars, about eight or more times the Sun's mass, die much more spectacularly than stars like our Sun. A **supernova** is a gigantic stellar explosion.

A massive star's carbon core contracts because of the force of gravity in the same way that a smaller star's does. But in the more massive star the core temperature continues to rise all the way up to 600 million K. At that point the carbon core begins to fuse. The collapse stops as carbon is converted into magnesium in nuclear fusion reactions.

When the carbon is used up, a new cycle is begun—gravitational contraction, rise in temperature, onset of new nuclear reactions, production of new elements, and a halt in the collapse. Elements heavier than carbon, such as nitrogen and silicon, are produced inside the star until the core is mainly iron.

Iron ends these cycles of nuclear fires and collapse because it does not release but instead requires energy in nuclear reactions. The doomed star collapses for the last time, until it cannot be compressed any further. Then it explodes violently. The light from the supernova can reach 100 billion times the Sun's luminosity. A supernova may be brighter than its whole galaxy for a short time.

Astronomers figure that most of the energy released in the explosion is invisible. A great amount is carried away at the speed of light by high-energy radiation and neutrinos ejected from the collapsing core. This energy holds clues to the causes of stellar explosions and the kinds and amounts of chemical elements manufactured and sprayed into space by supernovas.

Supernova 1987A, the first bright supernova in the sky since the telescope was invented, appeared in the large Magellanic Cloud in 1987. It was visible from the southern hemisphere for months and is the best-observed supernova to date (Figure 5.11). Neutrinos were detected exactly as theory predicted. The

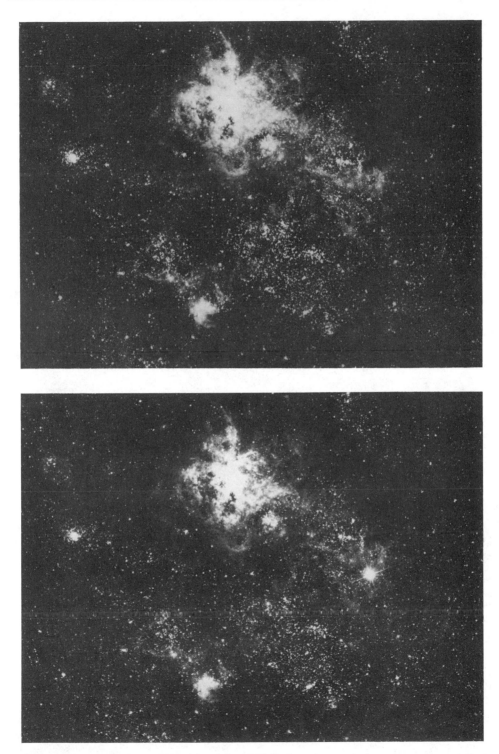

Figure 5.11. Supernova 1987A (top) in 1969 before the explosion was seen and (bottom) a week afterward, in February 1987.

core temperature at explosion must have been 200 billion K! Now astronomers are using Supernova 1987A data to refine and test theories of star death.

What kind of stars die as supernovas? _____

Answer: Very massive stars (about 8 or more times the Sun's mass).

5.15 SUPERNOVA REMNANTS

You might say that you are made of star dust.

Hydrogen and helium were probably the only elements in the universe when it began. Elements such as carbon, oxygen, and nitrogen, essential for life, are made inside the fiery cores of aging stars. The heaviest elements of all, such as gold and lead, are produced in the extremely high temperatures and intense neutron flux of a supernova explosion.

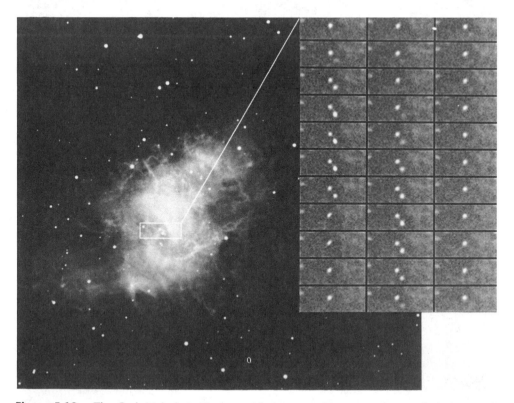

Figure 5.12. The Crab Nebula in Taurus, with the core of its progenitor exploded star still flashing at the center. The enlarged region is a mosaic of 33 one-millisecond slices in the period of the Crab Pulsar.

The supernova explosion sprays all these new elements out into space. They mix with the hydrogen, helium, and dust already there. All the material scattered into space by exploding massive stars becomes available again to be used in the formation of new stars and planets. Our Sun and Earth were formed about 5 billion years ago from a cloud of hydrogen and helium enriched in this way.

In A.D. 1054, Chinese and Native American observers recorded seeing a brilliant new star blaze in the sky even during daylight hours. The Crab Nebula in Taurus, a gas cloud expanding at 1600 km (1000 miles) per second, is observed today at the site of that supernova. It is about 3 pc (10 light-years) across, with the remnant core of the exploded star still at the center (Figure 5.12).

Which do you think are more abundant in the universe, elements lighter than iron or those heavier than iron? Why? _____

_____

_____

_____

_____

Answer: Lighter elements. These elements have much more time to form. Elements lighter than iron are produced from primordial hydrogen over a long period of time inside the cores of massive stars, while those that are heavier are produced only during the brief interval when the star explodes (supernova) at the end of its life.

5.16 SUPERDENSE STARS

When a very massive star explodes, it may leave behind a star of more mass than the Sun squeezed tightly together into a ball only about 16 km (10 miles) across. This extremely dense star is made mostly of **neutrons**, uncharged atomic particles. It was named a **neutron star** when it was first hypothesized.

Pulsars, pulsating radio stars, were first observed in 1967 by Jocelyn Bell, a graduate student at Cambridge University, England. Pulsars send sharp, strong bursts of radio waves to Earth with clocklike regularity, at intervals between milliseconds and 4 seconds. Hundreds of these strange objects have been observed so far (Figure 5.13).

Theory predicted that a neutron star should exist at the center of the Crab Nebula. A pulsar was found there in 1968 (Figure 5.12). The Crab Pulsar has since been observed over all electromagnetic wavelengths from radio to gamma.

A pulsar appears to be a rapidly rotating, highly magnetic neutron star.

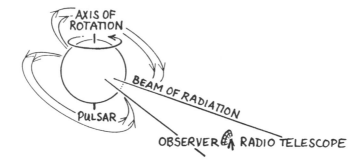

Figure 5.13. A pulsar or neutron star is too small to be seen. Instead, astronomers observe regular pulses of radiation beams emerging from the rotating star's magnetic poles as they sweep past Earth.

Its characteristic short, regular pulses come from radiation beams, emitted by very energetic accelerated charged particles, sweeping past Earth as the neutron star periodically spins. The rotation and pulse rates gradually slow down as energy is radiated away.

A pulsar, or neutron star, is the densest object observed to date.

How would you expect the force of gravity on the surface of a pulsar to compare to the force of gravity on Earth?

_____

_____

Answer: Much greater on a pulsar. The force of gravity is stronger the closer matter is packed, and a pulsar is extremely dense.

5.17 BLACK HOLES

A really massive star may continue to collapse after the pulsar stage to become a bizarre object called a **black hole** (Figure 5.14).

If black holes do exist, they are not holes at all. On the contrary, a black hole is a large mass contracted to extremely small size and enormous density. The force of gravity in such an object would be so great that, according to Einstein's theory of relativity, it would suck in all nearby matter and light.

A black hole can never be seen, because no light, matter, or signal of any kind can ever escape from its gravitational pull—hence its name. The surface of a black hole, or the boundary through which no light can get out, is called the **event horizon**.

The **Schwarzschild radius (R_S)** is the critical radius at which a spheri-

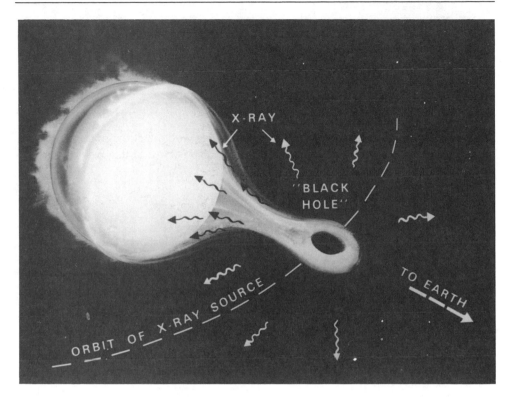

Figure 5.14. Artist's conception of a black hole.

cally symmetric massive body becomes a black hole. The equation is:

$$R_S = 2GM/c^2$$

where G is the gravitation constant, M is the mass of the body, and c is the speed of light (Appendix 2). The Schwarzschild radius for the Sun is about 3 km (2 miles) while for Earth it is about 1 cm (0.4 inch).

Theory predicts that a star of over three solar masses at its final collapse must cross its event horizon and disappear from view. No known force could stop further collapse, so the star may continue to shrink to a spot at the center called a **singularity**.

Cygnus X-1 is an intense X-ray source over 2500 pc (8000 light-years) distant in Cygnus. Discovered in 1966, it is an eclipsing binary star (period 5.6 days) whose unseen component is the first black hole candidate. The visible primary star is a blue supergiant that shows variations in spectral features from one night to the next. Possibly, when the unseen star sucks in material gravitationally from its visible companion, the observed X-rays are emitted.

You will surely hear more about these intriguing black holes in the future as scientists investigate them further.

What do you think would happen if an unlucky spaceship passed very close to a black hole in space? _____

_____

_____

Answer: The strong gravitational pull of the black hole would pull the spaceship in, producing a destructive force that would increase as the ship fell in and that would eventually tear it apart.

SELF-TEST

This self-test is designed to show you whether or not you have mastered the material in Chapter 5. Answer each question to the best of your ability. Correct answers and review instructions are given at the end of the test.

1. Define stellar evolution. _____

   _____

2. How do astronomers check a theory of stellar evolution? _____

   _____

   _____

   _____

3. List the three main steps in the birth of a star.

   (1)_____

   (2)_____

   (3)_____

4. What is the main source of the energy that a main sequence star shines into space? _____

   _____

5. For stars of the same initial chemical composition, what property determines the length of time it takes for the stars to evolve? _____

6. Why will the Sun stop shining as a main sequence star about 5 billion years from now? _____

   _____

   _____

7. List the seven main stages in the life cycle of a star like our Sun in order from birth to death.

   (1) _____   (2) _____

   (3) _____   (4) _____

   (5) _____   (6) _____

   (7) _____

8. List the seven main stages in the evolution of very massive stars in order from birth to death.

(1) _____ (2) _____

(3) _____ (4) _____

(5) _____ (6) _____

(7) _____

9. Why are elements that are lighter than iron, such as hydrogen, helium, carbon, and oxygen, so much more abundant in the universe than are the elements heavier than iron? _____

_____

_____

_____

_____

10. Match the eight items from the theory of stellar evolution to a real sky object.

____ (a) Birthplace of stars.   (1) Betelgeuse in Orion.
____ (b) Black hole candidate.   (2) Crab Nebula in Taurus.
____ (c) Blue giant.   (3) Crab pulsar in Taurus.
____ (d) Main sequence star.   (4) Cygnus X-1.
____ (e) Neutron star.   (5) Mira in Cetus.
____ (f) Pulsating variable star.   (6) Orion Nebula.
____ (g) Red giant.   (7) Rigel in Orion.
____ (h) Supernova remnant.   (8) Sun.

11. What is a black hole? _____

_____

_____

ANSWERS

Compare your answers to the questions on the self-test with the answers given below. If all of your answers are correct, you are ready to go on to the next chapter. If you missed any questions, review the sections indicated in parentheses following the answer. If you missed several questions, you should probably reread the entire chapter carefully.

1. The changes that take place in stars as they age—the life cycle of stars. (Section 5.1)

2. They predict what changes in luminosity and temperature should take place in stars as they age. Then they compare these theoretical tracks of evolution on H–R diagrams with H–R diagrams for groups of real stars.   (Section 5.1)

3. (1) Gravitational contraction of a cloud of gas and dust;   (2) rise in interior temperature and pressure;   (3) nuclear fusion.   (Section 5.3)

4. Nuclear fusion reactions in the core (hydrogen is converted into helium). (Sections 5.3, 5.5)

5. Mass.   (Section 5.4)

6. The Sun will leave the main sequence when all the available hydrogen fuel in its core is used up so that it no longer has an internal energy source. (Sections 5.6, 5.7)

7. (1) Protostar;   (2) main sequence star;   (3) red giant;   (4) variable star;   (5) planetary nebula ejected;   (6) white dwarf   (7) dead black dwarf. (Sections 5.3, 5.5 through 5.13)

8. (1) Protostar;   (2) main sequence;   (3) red giant;   (4) variable star;   (5) supernova;   (6) pulsar/neutron star;   (7) possible black hole (Sections 5.3, 5.5 through 5.7, 5.9, 5.14, 5.16, 5.17)

9. Hydrogen and some helium were probably the original elements in the universe. The other elements that are lighter than iron are formed inside aging stars over a period of time. Elements heavier than iron are formed only during the brief time of a supernova.   (Section 5.15)

10. (a) 6;   (b) 4;   (c) 7;   (d) 8;   (e) 3;   (f) 5;   (g) 1;   (h) 2. (Sections 5.2, 5.5 through 5.7, 5.9, 5.14 through 5.17)

11. A superdense, gravitationally collapsed mass from which no light, matter, or signal of any kind can escape.   (Section 5.17)

6

GALAXIES

*In nature's infinite book of secrecy
A little I can read.*

William Shakespeare (1564–1616)
Antony and Cleopatra, Act I, ii:11

Objectives

- ☆ Define a galaxy.
- ☆ Give the observational evidence for the Milky Way Galaxy's shape, size, structure, contents, and formation, and sketch the Galaxy showing the location of the Sun.
- ☆ Compare and contrast open (galactic) and globular clusters.
- ☆ Outline the method of using H–R diagrams to determine the ages of star clusters.
- ☆ Describe the contents of the interstellar medium.
- ☆ Compare and contrast emission and absorption nebulas.
- ☆ Explain how maps of our Galaxy in different wavelength regions are constructed.
- ☆ Identify the most distant object visible to the unaided eye.
- ☆ Compare and contrast the properties of spiral, elliptical, and irregular galaxies.
- ☆ Evaluate the evidence for two different models of galaxy formation and evolution.
- ☆ Define a cluster of galaxies and a supercluster.
- ☆ Give the observational evidence of large-scale structure in the universe.
- ☆ Compare and contrast the properties of a normal galaxy and active galaxies.
- ☆ Give the observed characteristics of quasars and a model that explains them.

6.1 STAR SYSTEMS

A **galaxy** is an enormous group of millions or billions of stars and gas and dust held together by the force of gravity.

Figure 6.1. A view toward the center of our Milky Way Galaxy in the constellation Sagittarius. Gas and dust clouds and myriad stars hide the galactic core from our sight.

Our Sun and all the visible stars in our sky belong to the **Milky Way Galaxy**. You may see a cloudy band of light across the sky on a very clear, dark night. The ancients named it the Milky Way because it looked like a trail of milk spilled in the sky by a goddess who was nursing her baby. That milky band is the combined glow of billions of stars in our huge Galaxy (Figure 6.1).

Try to locate the Milky Way overhead in summer or winter. If possible, use binoculars or a telescope to see that it is really made of many individual bright stars.

The entire Milky Way Galaxy contains over 200 billion stars. Those stars are very far apart from each other. On the average, a star's nearest neighbor star is about 5 light-years away.

What is a galaxy? _____

_____

Answer: An enormous collection of stars and gas and dust held together in space by the force of gravity.

Figure 6.2. Spiral galaxy M81 (NGC 3031) in Ursa Major resembles our Milky Way Galaxy. The arrow shows where our Sun and Earth would be if this were our Galaxy.

6.2 MILKY WAY GALAXY

Since we are bound to the Sun, which is located inside the huge Milky Way Galaxy, we cannot photograph our own Galaxy from the outside. Instead we use photographs of distant galaxies to help us picture what our own Galaxy must look like from space (Figure 6.2).

If you could go far into space and look down on our Galaxy, you would see a brilliant spiral pinwheel about 100,000 light-years (30 kpc) across. Our Earth, traveling around the Sun, is located out in one of the spiral arms.

If you could look at the Milky Way Galaxy from the side, it would look like a thin, shiny disk with a swollen center. The thickness of the central **nuclear bulge** is about 10,000 light-years (3 kpc). The thickness of the disk is about 3000 light-years (1 kpc). Our Sun is about 30,000 light-years (9 kpc) away from the center (Figure 6.3).

The whole Milky Way Galaxy is turning around in space. This fact is deduced from the Doppler shift of radiation from the spiral arms. Our Sun, with its family of planets, is racing around the center of our Galaxy at about 250 km/sec (563,000 miles per hour). Even at that incredible speed, our solar system requires about 220 million years to complete just one revolution.

Figure 6.3. Infrared image of the stars and gas clouds that form the disk of our Milky Way Galaxy, from the U.S. robot satellite Cosmic Background Explorer (COBE). Earth and Sun are at the outer edge of the disk, some 30,000 light-years from the Galaxy's center. The bright dots outside the disk are stars near the Sun. (The disk of the Galaxy extends far beyond our location.)

Our Galaxy appears to be hurtling through space in the direction of the constellation Hydra at a speed of over 600 km/sec (1 million miles per hour).

If you could fly across our Galaxy from one side to the other at light speed, how long would the trip take? _____

Answer: 100,000 years.

Solution: Divide the distance (100,000 light-years) by the speed (1 light-year per year).

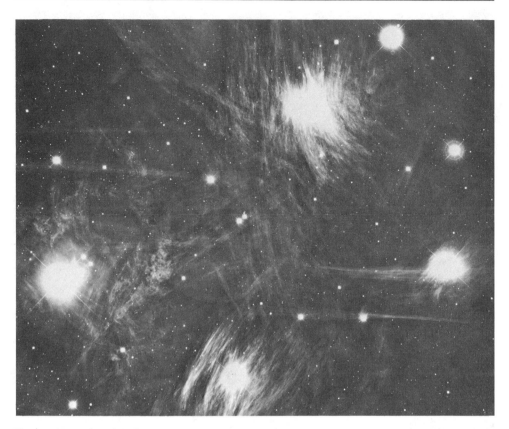

Figure 6.4 The Pleiades (M45) open star cluster in Taurus is visible to the eye as a group of six faint stars. It has hundreds of stars and is 400 light-years away. The glow around the stars is interstellar dust, which shines by reflecting starlight.

6.3 LOCATIONS OF STARS

Our Galaxy is a **spiral** type. Most of the stars are concentrated in a central **nucleus** and in spiral arms that wind out from it.

While some stars travel through the Galaxy alone, many move in **star clusters**, groups of stars that stay together because of their mutual gravitational attraction. Star clusters apparently form when a gigantic cloud of gas condenses into many stars. They are important to astronomers because all the different-mass stars in a cluster are about the same age. We see evidence of the cluster origin of stars in molecular clouds that contain hundreds of thousands of solar masses.

More than a thousand **open (galactic) clusters**, containing some 10 to 10,000 loosely packed stars each, have been observed. The stars move together in the disk. Open clusters are strongly concentrated in the spiral arms. Member stars are relatively young and typically hot and highly luminous (Figure 6.4).

A small fraction of the stars are in **globular clusters** in a **halo**, a spheri-

Figure 6.5. 47 Tucanae (NGC 104) in Tucana is the second brightest globular cluster. The core has a number of puzzling blue stragglers. Located 13,000 light-years away, 47 Tucanae looks to the eye like a fifth-magnitude star.

cal region around the disk. About 150 globular clusters, containing some 100,000 to 1 million tightly packed stars each, have been detected. They contain the oldest known stars (Figure 6.5).

Some globular clusters also have a small number of enigmatic **blue stragglers**, stars with an atypical blue color and high luminosity. They look much hotter and younger than the rest of the cluster's stars.

Refer to Table 6.1. List three differences between the open (galactic) clusters and the globular clusters found in our Galaxy. (1) _____

_____ ; (2) _____ ;

(3) _____

Answer: Open (galactic) clusters are found in the galactic disk, are relatively young, and have a smaller number of stars. Globular clusters are found in the galactic halo, are relatively old, and have a larger number of stars.

TABLE 6.1　Some Properties of Open and Globular Star Clusters

	Open (Galactic) Clusters	Globular Clusters
Location	Galactic disk	Galactic halo and nuclear bulge
Diameter	Under 100 ly	Over 100 ly
Age	Relatively young	Old
Number of stars	Up to 10,000	Up to 1 million
Color of brightest stars	Blue or red	Red

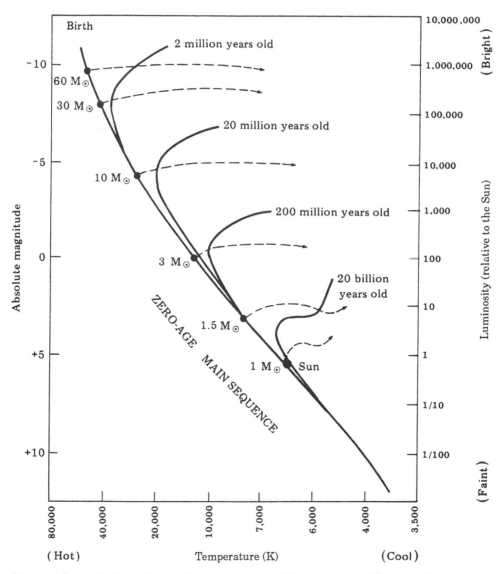

Figure 6.6. Solid lines give positions of stars in different clusters. The turnoff point away from the main sequence indicates the age of the cluster. Dashed lines are tracks of evolution of individual stars with masses shown. (M⊙= mass of Sun).

6.4 THEORY CHECK

Star clusters provide the best data for verifying a theory of stellar evolution.

First, H–R diagrams predicted by theory for stars of different ages are drawn. Then H–R diagrams for observed star clusters are drawn. The theoretical and observed diagrams are compared to verify or disprove the theory.

Figure 6.6 is a representation of predicted evolutionary tracks computed from theory. All stars start on the main sequence when they are born. The most massive stars are located at the top of the main sequence, and the least massive are at the bottom. All stars evolve away from the main sequence as

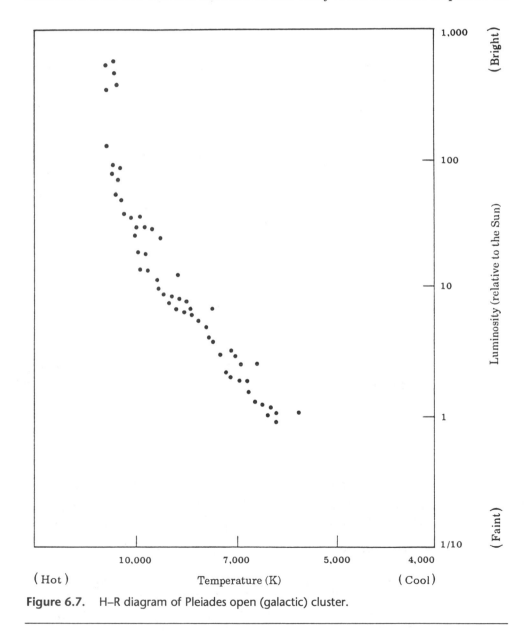

Figure 6.7. H–R diagram of Pleiades open (galactic) cluster.

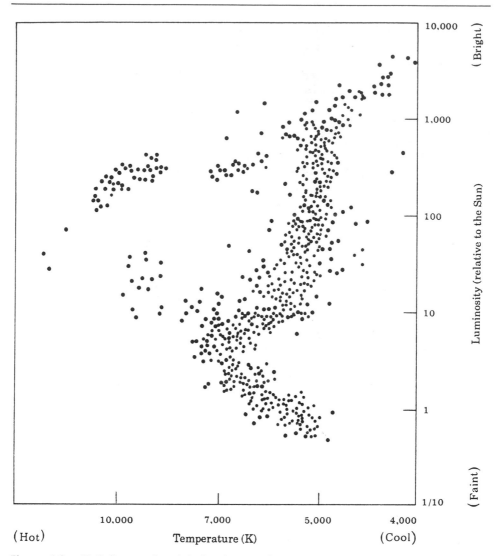

Figure 6.8. H–R diagram for globular cluster M3.

they age. Massive stars evolve fastest, so the higher the turnoff point, the younger the star cluster.

Compare the H–R diagrams for the Pleiades open cluster (Figure 6.7) and the M3 globular cluster (Figure 6.8) with the theoretical evolutionary tracks (Figure 6.6). State which is a relatively (a) young cluster _____ ; (b) old cluster _____. Explain your reasoning._____

_____

_____

Answer: (a) Pleiades cluster is relatively young. Most of its stars, even the massive short-lived ones, are still on the main sequence. (Pleiades cluster was born about 70 million years ago.) (b) M3 is relatively old. Hardly any stars appear on the upper half of the main sequence, and many stars have moved to the right into the red giant region. (M3 is about 10 billion years old.)

6.5 MASS

Until recently, the major part of the mass and luminosity of a **normal galaxy** such as ours was believed concentrated in stars like those near our Sun, whose mass is 2×10^{30} kg. Observed motions of stars and gas in galaxies now suggest that most of a galaxy's mass is in some still undetected form called **dark matter**. The visible galaxy may be surrounded by a much bigger and more massive, nonluminous **galactic halo**.

What would you expect the mass of the whole Milky Way Galaxy to be if it were concentrated in stars like our Sun? *Tip*: Use the approximate number of stars in our Galaxy from Section 6.1. _____

_____

Answer: Over 200 billion times the Sun's mass, or more than 4×10^{41} kg. (*Note*: The mass of the Milky Way Galaxy is at least 400 billion times the Sun's mass. It could be much more if dark matter exists.)

6.6 BETWEEN THE STARS

The space between the stars is practically empty on average, but local conditions vary a lot. The **interstellar medium**, matter and radiation between the stars, is mostly less dense than the air in vacuums produced on Earth.

Interstellar matter is particularly important because it is the raw material for new stars and planets. It is about 99 percent gas (about 75 percent of the mass of the gas is hydrogen and 23 percent is helium) and 1 percent **interstellar dust**, very tiny solid particles. In our Galaxy, most of the interstellar gas and dust is concentrated in the spiral arms, and that is where the newest stars are located.

Diverse clouds of gas and dust are continually enriched by material ejected by supernovas and stellar winds. An **H I region** is an intermediate-temperature cloud of neutral atomic hydrogen. An **H II region** is a cloud of ionized hydrogen near very hot stars.

More than 100 **interstellar molecules** besides hydrogen have been detected in dense, dark, cold, giant **molecular clouds**. Water vapor and common organic molecules are the most intriguing. These are key components of all known life on Earth. Their discovery in space has raised fascinating questions about the origin of life in the universe.

Why is it important in the theory of stellar evolution to know what interstellar matter consists of in any epoch? _____

_____

Answer: Interstellar matter is the raw material for new stars and planets.

6.7 GREAT CLOUDS

Historically, **nebula**, from the Latin for "cloud," was used for all kinds of hazy patches in the sky, including many now known to be star clusters or galaxies. The word is still sometimes used for a concentration of gas and dust.

A bright **emission nebula**, or H II region, is a cloud that glows by absorbing and then re-emitting starlight from very hot, young stars nearby. The Orion Nebula is a famous example you can observe (Figure 5.1).

Figure 6.9   The Horsehead Nebula (NGC 2024) in Orion is a famous dark nebula more than 1000 light-years away.

A dark **absorption nebula**, or molecular cloud, is a relatively dense concentration of interstellar matter whose dust absorbs or scatters starlight and hides stars that are behind it from our view.

Some nebulas are given fanciful names according to their appearances. What is the "horse's head" shown in Figure 6.9 actually made of? _____

_____

Answer: Relatively dense concentrations of interstellar dust.

6.8 MAPPING OUR GALAXY

We cannot look more than about a thousand light-years in most directions into our Milky Way Galaxy, even with the biggest optical telescopes, because dust clouds block our view.

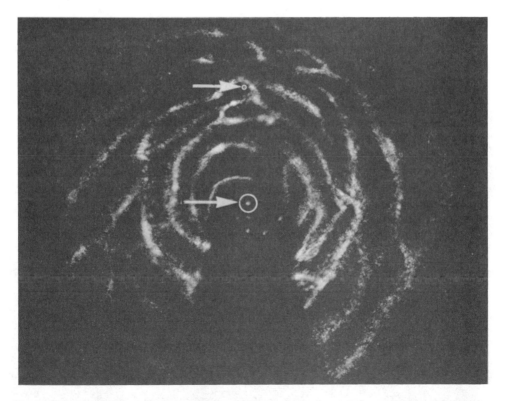

Figure 6.10.   A radio map showing the spiral structure of our Galaxy, produced by Leiden Observatory from observations of the 21-cm line. Large circle locates galactic center and small circle locates our solar system.

Astronomers use radio, infrared, and high-energy waves, which can pass through these clouds, to image the space beyond.

The spiral structure of our Galaxy is mapped by detecting radio waves of 21-centimeter wavelength. This **21-centimeter radiation** is emitted by neutral hydrogen atoms. It is strongest from regions with the biggest concentration of neutral hydrogen atoms—the spiral arms (Figure 6.10).

Large, hot gas clouds are mapped by detecting **continuous radio emission** rather than a particular wavelength. This continuous emission comes from concentrations of excited gas in hot H II regions.

Radio telescopes do not reveal exceptionally dense concentrations of hydrogen in dark, cool molecular clouds. In these regions hydrogen atoms join together to form hydrogen molecules. Radio astronomers map the densest gas concentrations by looking at a strong carbon monoxide emission line. Molecular hydrogen, which does not emit or absorb radio radiation, is observed at infrared and ultraviolet wavelengths.

New data continually modify our views. Particularly intriguing observations include stellar coronas and very hot intercloud gas at ultraviolet wavelengths, as well as **X-ray bursters**, X-ray stars showing violent random changes in their emissions, and **gamma ray bursters**, sources of transient bursts of gamma radiation.

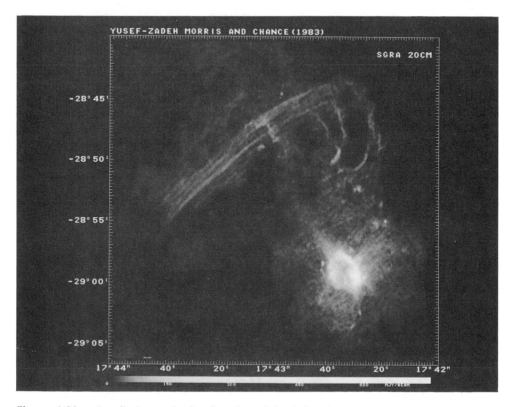

Figure 6.11. A radio image in the direction of the Galaxy's center.

The nucleus of our Galaxy apparently contains an extraordinary, very massive, compact object ringed by very hot, chaotic gas clouds and dust. Possibly a massive black hole or dense, luminous star cluster powers the central gas flows and luminosity. As matter falls in toward the center, it is compressed and heated to millions of degrees, producing the observed X-rays (Figure 6.11).

If, as recent observations indicate, our Galaxy has a barred spiral structure, the rate of infall would be much more rapid than in normal spirals. Violent central starbursts, in which great numbers of very bright and massive stars form, would occur.

What is particularly interesting about regions of relatively dense gas concentrations in our Galaxy? _____

Answer: Stars are forming in these regions.

6.9 STAR POPULATIONS

In 1944, U.S. astronomer Walter Baade (1893–1960) divided stars into two classes. Although now known to be oversimplified, this classification was useful for first explaining how age, dynamics, and element production in stars and galaxies are related.

Population I stars include the hottest and most luminous stars. These relatively young stars are located in the disk, especially in the spiral arms, embedded in the dust and gases from which they formed. They are relatively high in heavy elements (similar to the Sun, about 1 percent by mass) in addition to their hydrogen and helium.

Population II stars, like those in globular clusters, are found toward the galactic nucleus and in the halo. These stars are older. They are made almost entirely of hydrogen and helium.

How does the stellar evolution theory explain the difference between Population I and Population II stars? _____

_____

_____

_____

Answer: Population II stars are the oldest stars. They formed out of the original hydrogen and helium that was available when the Milky Way Galaxy was born. Population I stars are young. They formed much later out of the dust and gas in space that was enriched by elements manufactured in stars and sprayed into space by supernovas.

6.10 FORMATION OF OUR GALAXY

Our Galaxy appears to have formed 10 to 20 billion years ago, perhaps a few hundred million years after the universe began. The oldest stars are about 13 to 18 billion years old.

A popular model says our Galaxy originated in a turbulent, rotating cosmic cloud of hydrogen and helium. The cosmic cloud collapsed to a coherent structure when the inward pull of gravity finally exceeded the outward pressure. Forces of gas pressure, radiation, rotation, and gravity then shaped our Galaxy to its current form.

What would this model of galaxy formation predict (a) the oldest and (b) the youngest stars in our Galaxy to be made of? Explain your answer. _____

_____

_____

_____

_____

Answer: (a) Hydrogen and helium, the elements present as raw materials at the time our Galaxy was new. (b) Hydrogen, helium, and the other 90 naturally occurring elements. The interstellar medium is the raw material of new stars. Originally consisting of hydrogen and helium, it has been enriched by elements expelled by supernovas and stellar winds.

6.11 STRUCTURE OF THE MILKY WAY GALAXY

Refer to Figure 6.12. As a summarizing activity with regard to our Milky Way Galaxy, identify the following: (a) disk _____ ; (b) halo _____ ; (c) spiral arm _____ ; (d) nucleus _____ ; (e) position of Sun and Earth _____ ; (f) location of globular clusters _____ ; (g) nuclear bulge

Answer: (a)2; (b)1; (c)3; (d)4; (e)6; (f)5; (g)7.

6.12 BEYOND THE MILKY WAY GALAXY ✪

Our Galaxy was the only one recognized until 1924. Then U.S. astronomer Edwin Hubble (1889–1953) analyzed Cepheid variables and proved that some of the fuzzy "nebulas" previously observed were really distant galaxies.

The **New General Catalog (NGC)** of nonstellar astronomical objects was

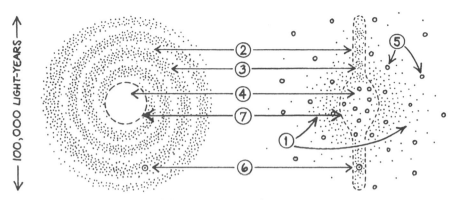

Figure 6.12. Two views of the Milky Way Galaxy.

first published in 1888 as a list of 7840 nebular objects compiled by Danish astronomer Johann Dreyer (1852–1926). It was expanded in 1895 in a supplement, the **Index Catalog (IC)**, and again in 1908 in the **Second Index Catalog**. The **Messier Catalog** of 110 nebulas, star clusters, and galaxies (Appendix 6) was originally a list of 45 hazy objects compiled in 1784 by French astronomer Charles Messier (1730–1817), who wanted to avoid mistaking them for new comets.

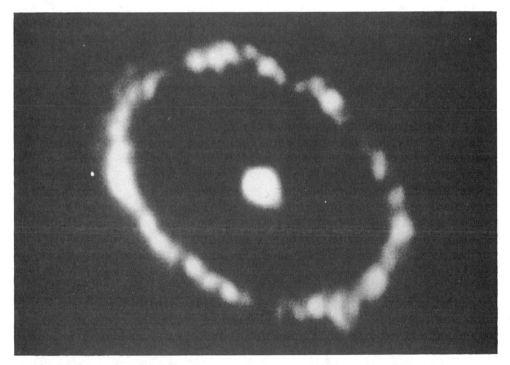

Figure 6.13. Supernova 1987A (at the center) and Ring in the Large Magellanic Cloud in the constellation Tucana.

Today we believe that the universe is full of galaxies, perhaps 100 billion of them, having typically over 100 billion stars each. Most bright galaxies are identified by an NGC, IC, or M (Messier catalog) number.

Two small, irregularly shaped galaxies, the Large Magellanic Cloud (LMC) in the constellation Tucana and the Small Magellanic Cloud (SMC) in Dorado are nearby satellite galaxies of our Milky Way. They are held by the force of gravity at a distance of about 169,000 light-years (52 kpc) and 210,000 light-years (60 kpc), respectively. The distance to the LMC was deduced to within 5 percent from observations of the elliptical ring around Supernova 1987A (Figure 6.13), which was ejected by the progenitor star when it was a red supergiant.

The Magellanic Clouds are visible to the unaided eye from the southern hemisphere and were first noted by Portuguese explorer Ferdinand Magellan (c. 1480–1521) on his historic trip around the world. Both belong to a system that is inside an immense, invisible hydrogen envelope detected at 21-cm wavelength.

The Andromeda Galaxy (M31, NGC 224) is the closest galaxy that is similar to ours, with perhaps twice the mass (Figure 6.14). It is the most distant object, about 2.5 million light-years (670 kpc) away, that you can see with

Figure 6.14. Andromeda Galaxy with bright companions NGC205 above and M32 below, photographed through a 1.2-m (48-inch) telescope. The spiral Andromeda Galaxy is even bigger than our Galaxy and contains billions of individual stars.

your unaided eye and a special sight in small telescopes. In the fall, look for a fuzzy patch of light in the Andromeda constellation (shown as O GALAXY on your star maps).

To appreciate what "close" means for galaxies, estimate how many Milky Way Galaxies you could line up next to each other between us and our neighbor, Andromeda Galaxy. _____

Answer: 25.

Solution: $\dfrac{\text{Distance to Andromeda Galaxy}}{\text{Diameter of Milky Way Galaxy}} = \dfrac{2,500,000 \text{ ly}}{100,000 \text{ ly}} = 25$

6.13 CLASSIFICATION OF GALAXIES

Galaxies come in several different shapes and sizes. They were first classified into groups according to their structure by Edwin Hubble in 1926 (Figure 6.15).

Elliptical galaxies, designated E, are egg shaped. They range from nearly perfect spheres, E0, to the flattest, E7. Elliptical galaxies seem to contain practically all old stars. They have little visible gas and dust, but infrared and X-ray observations reveal some.

Spiral galaxies are divided into two major subcategories. **Normal spiral galaxies**, S, have a bright disk where spiral arms wind out from a bulging nucleus. They are subdivided into Sa, Sb, and Sc, according to the size of the

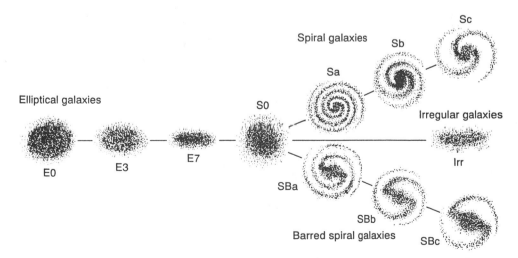

Figure 6.15. Hubble classification of galaxies according to shape and arranged in a tuning fork diagram.

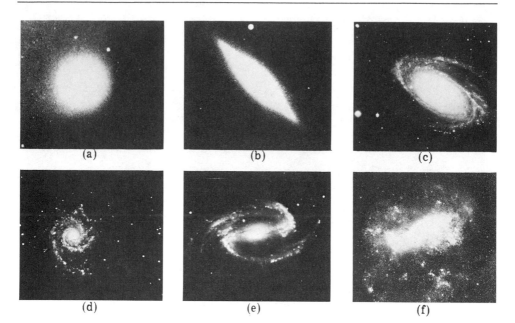

Figure 6.16. Galaxies of various classifications.

central bulge and to how tightly wound the spiral arms are. Those with bright flat disks but no spiral arms are designated SO. **Barred spiral galaxies**, SB, look like normal spiral galaxies except that the spiral arms unwind from the ends of a bar-shaped concentration of material. Spiral galaxies have large amounts of gas and dust in the disk and contain young, middle-aged, and old stars.

Irregular galaxies, Ir, have no regular geometric shape. They usually contain gas and dust, mostly bright young stars and clouds of ionized gas, and some old stars.

Hubble began the systematic study of distant galaxies using the 2.5-m (100-inch) Mount Wilson telescope, the world's largest, from 1918 to 1938. Today astronomers are gathering a dizzying variety of data using radio, infrared, ultraviolet, X-ray, and gamma ray detectors in addition to the giant optical telescopes.

Classify each of the galaxies in Figure 6.16 by its shape. (a) _____ ; (b) _____ ; (c) _____ ; (d) _____ ; (e) _____ ; (f) _____

Answer: (a) E0; (b) E7; (c) Sb; (d) Sc; (e) SBb; (f) Ir.

6.14 GALACTIC PROPERTIES

The distance to a galaxy is a key to determining the galaxy's basic properties. Distance measurements are difficult to make and are still uncertain. The uncertainties carry over to the determination of other galactic data.

A **standard candle**, or astronomical object whose absolute magnitude is known from its observed characteristics, can be used to determine distances to galaxies out to about 10 million light-years. Useful standard candles include Cepheid variables, the most luminous stars, globular clusters, supernovas, and standard types of galaxies.

Standard candles are hard to calibrate. A new technique uses a correlation between spectral line width and luminosity of a spiral galaxy to determine absolute magnitude and distance.

When a galaxy's distance is determined, its diameter and luminosity can be figured from its apparent magnitude and apparent diameter.

The mass of a galaxy is calculated from observed gravitational effects on stars or gas clouds within it or on neighbor galaxies. Observational data indicate that most of a galaxy's mass is unobserved dark matter. Apparently a large spiral or elliptical galaxy is a trillion times more massive than our Sun.

Refer to Table 6.2, which summarizes rough values of the data collected so far. (Values for individual galaxies may vary a lot.) State two differences between spiral and elliptical galaxies.

_____

_____

_____

_____

Answer: Spirals contain both old and young stars; they have visible gas and dust between the stars to make new stars. Ellipticals contain old stars; they have little visible interstellar gas and dust.

TABLE 6.2   Rough Values of Galactic Data

Values	Spirals	Ellipticals	Irregulars
Mass (Milky Way Galaxy = 1)	0.005–2	0.000001–50	0.0005–0.15
Diameter (Milky Way Galaxy = 1)	0.2–1.5	0.01–5	0.05–0.25
Luminosity (Milky Way Galaxy = 1)	0.005–10	0.00005–5	0.00005–0.1
Population content of stars	Old and young	Old	Old and young
Luminous inter-stellar matter	Moderate gas and dust	Little gas and dust	Plentiful gas and dust

6.15 GALACTIC EVOLUTION

Many mysteries about galaxies still challenge astronomers.

Which formed first, galaxies or stars? Although a galaxy changes much too slowly for a human to watch, astronomers can see galaxies at ever earlier stages of development by observing them at ever greater distances from us. (Looking farther out into space is seeing further back in time.) If galaxies formed from primordial material in the early universe, observers should detect a **protogalaxy**, or galaxy in formation.

After birth, does a galaxy evolve gradually by accumulating nearby gas, or does it reach its final size quickly? Inside a galaxy, stars are born, evolve, and die and return matter enriched with heavy elements to space for the formation of new stars. Galaxies may also evolve dynamically by interacting with each other. Large elliptical galaxies may result from slow collisions of smaller spirals. A collision could create a single system with merged nuclei surrounded by a halo of stars.

When does a galaxy assume its shape? A galaxy's shape may be determined mainly by its initial mass, density, and angular momentum and by whether it has close companions or belongs to a cluster. Or spiral galaxies may form preferentially first, and elliptical galaxies form later by mergers of spiral galaxies.

How do the chemical composition, color, and luminosity of a galaxy change over billions of years? Color and luminosity changes with age were confirmed recently by comparing galaxies some 10 billion light-years away (younger) with nearby (older) galaxies. The more distant, younger galaxies were brighter and bluer. Apparently hot blue stars form at a higher rate in young galaxies than in old galaxies.

Did many galaxies go through an extremely energetic early stage? Many more active galaxies are at great distances than nearby. Perhaps many galaxies went through an extremely energetic early stage in which **quasars**, small, extraordinarily luminous extragalactic objects with high redshifts, were their energy-producing centers.

Refer to Figure 6.15 and Table 6.2. What observed data indicate that the different shapes of galaxies do not represent stages of evolution in the life cycle of a galaxy?_____

_____

_____

_____

Answer: All three types of galaxies contain old stars. That fact indicates that spiral and irregular galaxies are just as old as elliptical galaxies and could not be the end stage of a galaxy's life. Nor could elliptical galaxies be the first stage of a life cycle, as proposed by Hubble, because they do not have the dust and gas necessary for the birth of new stars seen in spiral and irregular galaxies.

6.16 GROUPINGS

Photographic surveys of the sky show that most galaxies belong to groups, called **clusters of galaxies**. These clusters contain from several to thousands of galaxies held together by the force of gravity as they orbit one another at velocities of about 1000 km (600 miles) per second. **Richness** denotes the number of galaxies above a selected brightness level within a cluster. **Structure** refers to a grouping of galaxies (Figure 6.17).

Our Milky Way Galaxy belongs to a typical small cluster, the **Local Group**, with about 30 members. "Local" means that the galaxies are within a region 3 million light-years across. Three of these galaxies—our Milky Way, Andromeda (M31), and M33 in Triangulum—are spirals. The others are ellipticals (including M31's bright companions NGC 205 and M32) or irregulars (including the Magellanic Clouds). Several are **dwarf galaxies**, small, low-mass galaxies a few thousand light-years in diameter.

Clusters divide into two classes by shape. **Regular clusters** are relatively compact, with highest density near the center. Members are mostly elliptical and SO galaxies. Many regular clusters emit radio radiation from active galax-

Figure 6.17. The Virgo cluster includes thousands of galaxies with the Local Group on its periphery. It is the nearest rich cluster of galaxies, about 50 million light-years away.

ies and **intergalactic** (between the galaxies) **gas**. About a third emit X-rays from intergalactic gas at about 100 million K.

In contrast, **irregular clusters**, including our Local Group, have a looser structure with little central concentration and less very hot gas. They contain many spiral and irregular galaxies. Fewer emit radio waves or X-rays.

A **supercluster** is a cluster of clusters of galaxies. Superclusters are the largest gravitationally bound systems observed so far. They are about 100 million to 1 billion light-years across. The Local Group with our Milky Way Galaxy belongs to the Virgo Cluster, which is a part of the **Local Supercluster**.

Superclusters are located in thin sheets that border **voids**, regions where few galaxies are observed. The voids are like gigantic bubbles with clusters of galaxies along their surfaces. The observable universe consists mostly of vast voids between superclusters.

What is the largest type of structure in the universe? _____

Answer: Supercluster of galaxies.

6.17 EXTRAORDINARY ACTIVITY IN GALAXIES

Some galaxies have strange forms and unusual characteristics.

An **active galaxy** is a galaxy whose center, or **active galactic nucleus (AGN)**, emits exceptionally large amounts of energy. This energy far exceeds

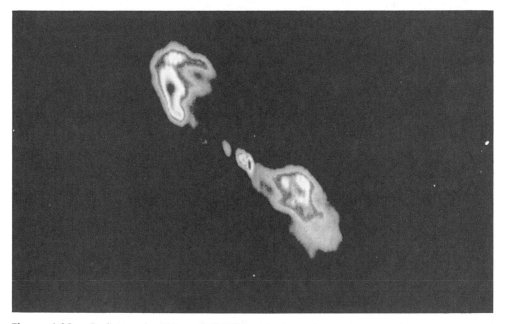

Figure 6.18.   Radiograph of Virgo A (M87), an active galaxy.

the total output from normal nuclear fusion reactions in stars in normal galaxies. AGNs often have great jets of hydrogen gas racing outward at very high speeds (Figure 6.18).

The source of the colossal energy seems to be a central, powerful source of gravity that sucks in nearby matter. A very massive object, such as a black hole with a mass millions of times that of the Sun, could be the attracting force. In this model, as dust, gas, and even stars spiral in toward the black hole they accelerate and heat up. Fiery, dense, infalling matter emits the radiation. The mass of the attractor may be inferred from the speed of infall.

What is a likely explanation of the violent activity in the galaxy shown in Figure 6.18? _____

Answer: A black hole at the galaxy's center.

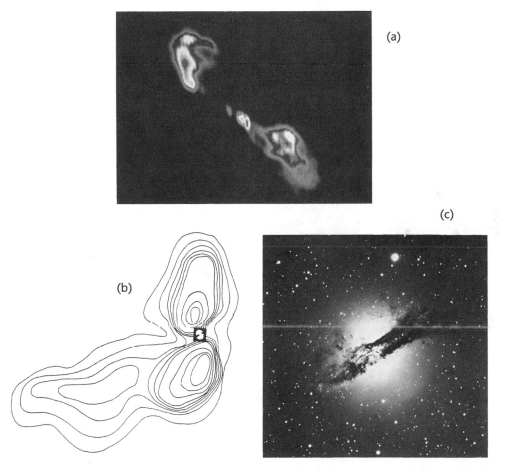

Figure 6.19. The (a) radiograph, (b) radio contour map, and (c) visible image of Centaurus A, a double-lobed radio galaxy with jet.

6.18 RADIO GALAXIES

Radio galaxies are the largest class of active galaxies (Figure 6.19).

The radiograph of a typical radio galaxy shows two large patches of energy at radio wavelengths on opposite sides of a visible galaxy. The radio energy usually looks like **synchrotron radiation**, or radiation produced by electrons spiraling around at nearly the speed of light in a strong magnetic field.

If the hypothetical central black hole exists, the jets of high-speed electrons are ejected by matter as it disappears into the black hole. The electrons emit the colossal radio energy while accelerating in a strong magnetic field.

Galactic close encounters could provide the prodigious matter that the hypothetical black hole devours.

Apparent collisions between galaxies and cases of galaxies passing close to each other have been photographed. When two galaxies collide, they apparently pass through each other. The clouds of gas and dust in **colliding galaxies** would be unusually dense and could trigger a burst of star formation or fuel the hypothetical black hole (Figure 6.20).

Galactic cannibalism probably occurs when a very large galaxy passes too close to a much smaller one and "eats" it. A massive galaxy could tidally strip away and take in gas, dust, and stars from a smaller disk. The smaller nucleus, falling to the center of the massive galaxy, would fuel the larger's energy output for millions of years.

Figure 6.20. Colliding galaxies NGC 4676A/NGC 4676B, "The Mice." Long tails of stars trailing away from the central regions of the galaxies result from the interaction.

What do you think might happen to life on Earth if our Galaxy collided with another galaxy? Explain your answer. _____

_____

_____

Answer: Probably nothing. The stars and their possible planets are separated by such vast distances inside galaxies that two galaxies can pass through each other without their stars ever coming into contact with each other. (No star collisions have ever been observed.)

6.19 SEYFERT GALAXIES

A **Seyfert galaxy**, named for U.S. astronomer Carl K. Seyfert (1911–1960), who described the prototype, is a spiral galaxy with an active galactic nucleus (Figure 6.21).

Figure 6.21.   Seyfert galaxy Perseus A (NGC 1275) is a radio and X-ray source. An extensive system of long filaments is material exploding outward into space at 2500 km (1500 miles) per second.

A **Seyfert nucleus**, typically only about 10 light-years across, shines many times more brilliantly than a normal galaxy the size of our Milky Way. Its spectrum has broad emission lines that indicate turbulent motions of very hot gas at velocities of thousands of kilometers per second.

Most Seyfert galaxies are strong emitters of infrared radiation. Heated dust enveloping the nucleus probably absorbs high-energy radiation emitted from the energized core and re-emits it at longer infrared wavelengths.

Less than 2 percent of all spiral galaxies are Seyferts. Either all spirals have active nuclei at some time, or only a small fraction of spirals act up in this way.

How is a Seyfert galaxy different from a normal spiral galaxy? _____

_____

_____

Answer: A Seyfert galaxy has a small, exceptionally brilliant nucleus with broad emission lines (that do not come from stars) in its spectrum.

6.20 MYSTERIOUS QUASARS

The first quasars observed look like faint stars in photographs but are strong radio sources with nonstellar spectra, hence the name quasi-stellar radio source and its contraction to quasar (Figure 6.22).

Most of the thousands of quasars emit extraordinary power across a

(a)   (b)

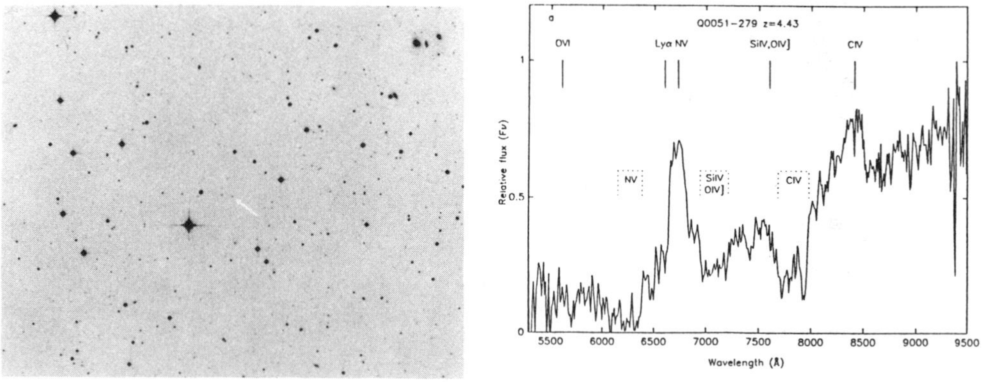

Figure 6.22.   Extremely distant Quasar Q0051-279 (a) is practically indistinguishable from stars in ordinary photographs. (b) Its spectrum has light so far shifted to the red as to place this quasar near the assumed beginning of the universe.

broad range of wavelengths, from radio to gamma rays, but the original name is kept. Perhaps radio emission is a temporary phase in their evolution.

Quasars are small for celestial objects, typically about 1 light-day across (not much bigger than our solar system), but they may shine brighter than a thousand normal galaxies. Almost all quasars vary irregularly in their light output.

The light from quasars is highly shifted toward the red end of the spectrum. Quasars have the highest redshifts observed. Most astronomers interpret this quality as a Doppler shift, meaning the quasars are racing away from us at speeds of over 90 percent the speed of light. If the interpretation is accurate, quasars are the most distant and most powerful objects ever detected.

Ultraviolet light emitted by a quasar with the largest redshift is received as red light on Earth. If indeed this effect is a Doppler redshift, the edge of the known universe is marked by quasars racing away at fantastic speeds of over 1 billion km (600 million miles) per hour. These quasars were shining when the universe was young.

Twin or multiple images of the same apparent quasar support the view

Figure 6.23. Gravitational Lens G2237 + 0305, the "Einstein Cross," from the U.S. Hubble Space Telescope. Light from a quasar about 8 billion light-years away is bent in its path by the gravitational field of a galaxy (diffuse central object) at a distance of 400 million light-years to form the four bright outer images.

that quasars are located at cosmological distances. According to Einstein's theory of general relativity, starlight passing near a massive body is deflected. A galaxy much closer to us than a particular quasar could be a **gravitational lens** that produces multiple images of the quasar (Figure 6.23).

Different hypotheses have been proposed and abandoned to explain the stupendous energy output of these cosmic powerhouses. Einstein's theory of relativity was used to attribute the quasars' extraordinary redshift to an enormous gravitational force **(a gravitational redshift)**, which would have meant the quasars were closer and not so powerful after all. Collisions of material particles with **antimatter**, exotic opposite of matter on Earth, were invoked. A new, still-unknown source of energy was suggested.

CCD detectors show brilliant, compact quasars at the centers of galaxies. A quasar may be the most luminous type of active galactic nucleus. Since quasar activity was much more common in the early universe than it is today, a quasar could be a development phase in young galaxies.

Astronomers are cataloguing and analyzing the sizes, shapes, brightnesses, colors, redshifts, and distributions of numerous galaxies and quasars to gain more understanding of the universe.

What is mysterious about the quasars? _____

_____

_____

Answer: The source of the stupendous energy output they must have if they are really so extremely far away as believed.

SELF-TEST

This self-test is designed to show you whether or not you have mastered the material in Chapter 6. Answer each question to the best of your ability. Correct answers and review instructions are given at the end of the test.

1. Define a galaxy. _____

   _____

2. Arrange the following in order of increasing size: star, planet, galaxy, cluster of galaxies, open cluster, supercluster, solar system. _____

   _____

3. Sketch an edge-on view of the Milky Way Galaxy, and label the (a) size of the diameter; (b) disk; (c) nucleus; (d) spiral arm; (e) halo; (f) position of Sun and Earth; (g) location of globular clusters. _____

4. Which of the following have been identified in the interstellar medium: hydrogen gas, radiation, bacteria, tiny solid dust particles, viruses, water vapor, spirits, gases of elements heavier than hydrogen, organic molecules, algae? _____

   _____

5. Why is it important in the theory of stellar evolution to know what interstellar matter consists of in any epoch?_____

   _____

6. Refer to Figure 6.9. Is the space behind the "horse's head" really empty of stars? Explain._____

   _____

   _____

7. Why is the 21-cm radio radiation emitted by hydrogen atoms more useful than visible light in mapping the structure of our Milky Way Galaxy?

   _____

   _____

8. Refer to the H–R diagrams of star clusters 1 and 2 in Figures 6.24(a) and (b).

(a) Which cluster is older? _____

(b) Which cluster would have population I stars? _____

(c) Which cluster would have stars with a relatively high abundance of heavy elements?_____

(d) Which is a globular cluster? _____

(e) Which has many bright blue stars? _____

(f) Which may contain up to 10 million stars?_____

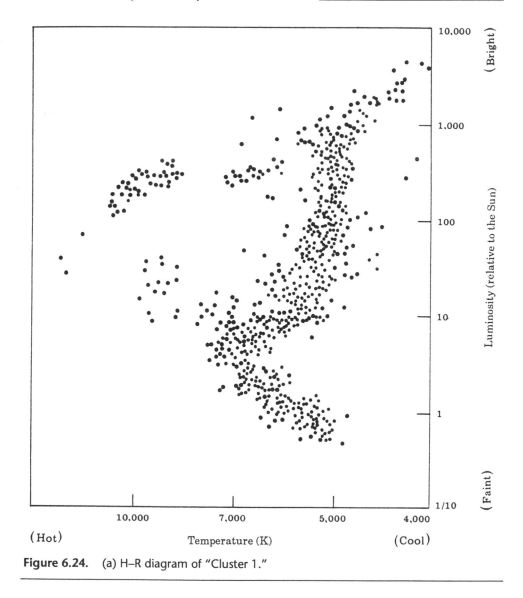

Figure 6.24.   (a) H–R diagram of "Cluster 1."

9. (a) What is the most distant object visible to the unaided eye? _____

_____

_____

(b) How long does it take light emitted from that object to reach your eyes?

_____

_____

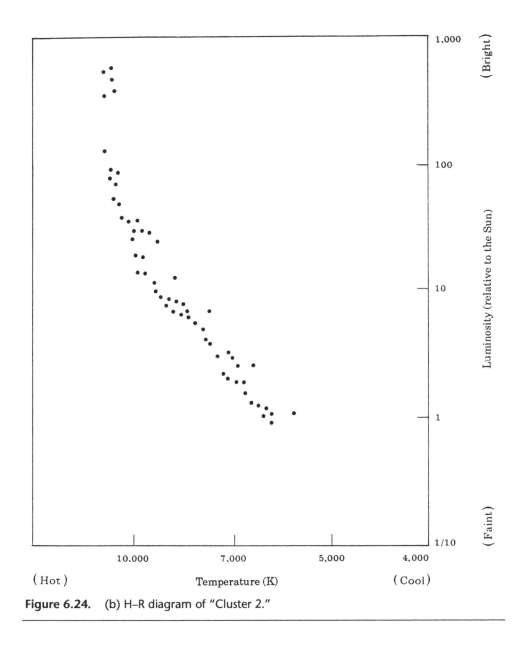

Figure 6.24. (b) H–R diagram of "Cluster 2."

10. List the main shapes of galaxies in the Hubble classification scheme. and explain why they cannot represent successive stages of galaxies' evolution.

_____

_____

11. What is the most popular explanation for the colossal energy output of active galaxies? _____

_____

12. Match the following descriptions with the correct object:

____ (a) Shows the largest redshift known.

____ (b) The clouds of gas and dust here are much more dense.

____ (c) Radiograph shows two large patches emitting radio waves on opposite sides of a visible galaxy located between them.

____ (d) Has a relatively small brilliant nucleus with broad emission lines in its spectrum.

____ (e) Its luminosity can be explained as the composite of a collection of many individual stars.

(1) Colliding galaxies.
(2) Seyfert galaxy.
(3) Normal galaxy.
(4) Quasar.
(5) Radio galaxy.

ANSWERS

Compare your answers to the questions on the self-test with the answers given below. If all of your answers are correct, you are ready to go on to the next chapter. If you missed any questions, review the sections indicated in parentheses following the answer. If you missed several questions, you should probably reread the entire chapter carefully.

1. An enormous collection of stars and gas and dust held together in space by the force of gravity.   (Section 6.1)

2. Planet, star, solar system, open cluster, galaxy, cluster of galaxies, supercluster.   (Sections 6.1 through 6.3, 6.16)

3. Figure 6.25.   (Sections 6.2, 6.3, 6.11)

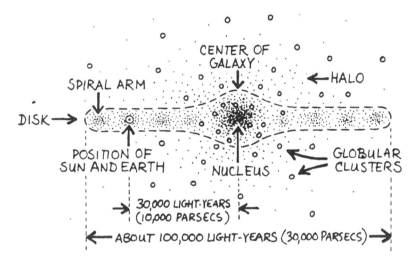

Figure 6.25.   Edge-on view of the Milky Way.

4. Hydrogen gas, radiation, tiny solid dust particles, water vapor, gases of elements heavier than hydrogen, organic molecules.   (Section 6.6)

5. Interstellar matter is the raw material for new stars and planets.   (Sections 6.6, 6.8)

6. No. The "horse's head" is a dark nebula. It is a relatively dense concentration of interstellar matter whose dust absorbs or scatters starlight and hides stars that are behind it from our view.   (Section 6.7)

7. Radio waves pass through the interstellar dust in the disk of the Milky Way Galaxy much more effectively than visible light waves.   (Section 6.8)

8. (a) 1; (b) 2; (c) 2; (d) 1; (e) 2; (f) 1. (Sections 6.3, 6.4, 6.9)

9. (a) Andromeda Galaxy; (b) around 2.2 million years. (Section 6.12)

10. Elliptical, spiral, irregular. All contain old stars, so all must be equally old. (Sections 6.13 through 6.15)

11. A very massive object, probably a black hole, at the galaxy's center. (Section 6.17)

12. (a) 4; (b) 1; (c) 5; (d) 2; (e) 3. (Sections 6.5, 6.17 through 6.20)

7

THE UNIVERSE

In the beginning God created the heaven and the Earth. Now the Earth was unformed and void, and darkness was upon the face of the deep; and the spirit of God hovered over the face of the waters. And God said: "Let there be light." And there was light. And God saw the light, that it was good.

Genesis 1:1–4

Objectives

☆ Define cosmology.

☆ Describe the basic assumptions and limitations of cosmology.

☆ Specify the evidence that the universe is expanding.

☆ State the Hubble law.

☆ Explain the significance of the Hubble constant.

☆ Describe the past and present of the universe according to the Big Bang theory.

☆ Compare and contrast the future of the universe according to the open, flat, and closed models of the universe.

☆ List important observations in support of the Big Bang theory.

☆ Describe methods for choosing among the open, flat, and closed models of the universe.

☆ Outline a problem of the standard Big Bang model and its resolution by the inflationary universe model.

☆ Describe astronomical methods of estimating the age and size of the universe.

7.1 ETERNAL QUESTIONS

People have always wondered about how the world began and if it will end. Ancient myths, philosophy, and theology all provide models. **Cosmology** is the study of the origin, present structure, evolution, and destiny of the universe.

Astronomers construct **cosmological models**, mathematical descriptions that try to explain how the universe began, how it is changing as time goes by, and what will happen to it in the future. These models must be consistent with the observational data we have on stars and galaxies.

Two basic types of models, **evolutionary** and **steady state,** have been tested in the last 50 years. Results substantiate the evolutionary type.

Cosmological models differ from religious explanations of the universe in a fundamental way. Can you state the difference? _____

Answer: Cosmological models do not give a supernatural cause or meaning to physical events but try to explain these events using only the laws of nature and mathematics.

7.2 THE EXPANDING UNIVERSE

The basic observation that must be accounted for by any cosmological model is that light from distant galaxies is shifted in wavelength toward the red end (longer wavelengths) of the spectrum. This phenomenon is called the **cosmological redshift**.

Modern theory says this redshift is a Doppler effect (Section 3.9), indicating that other galaxies are racing away from us. The most distant galaxies we observe have the greatest redshifts. They are moving away fastest of all (Figure 7.1).

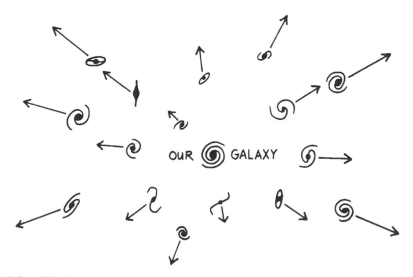

Figure 7.1 Galaxies receding, as they appear from our Milky Way Galaxy (arrows indicate velocities).

When we look into space we see galaxies receding from us. What does this observation imply about the universe? _____

_____

Answer: The universe must be expanding.

7.3 REDSHIFTS

Examine Figure 7.2, which shows redshifts and corresponding calculated velocities for five galaxies that are at different distances from us.

Laboratory spectral lines of known wavelength are shown above and below the spectral lines of each galaxy for reference. A pair of the darkest

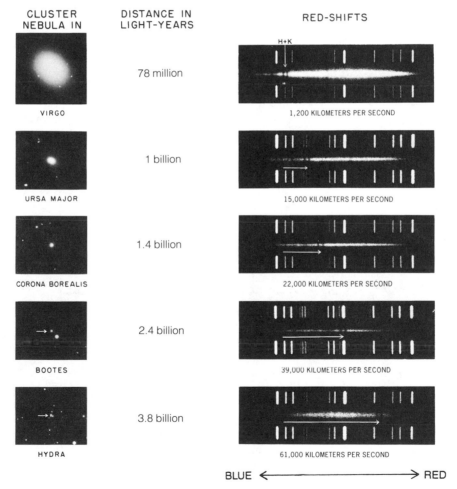

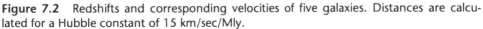

Figure 7.2   Redshifts and corresponding velocities of five galaxies. Distances are calculated for a Hubble constant of 15 km/sec/Mly.

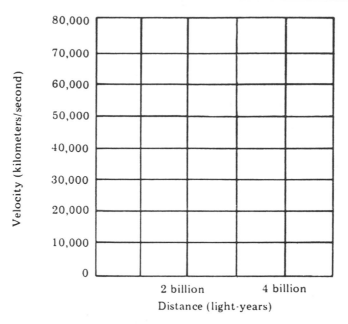

Figure 7.3a Grid for velocity versus distance diagram.

absorption lines, H and K of ionized calcium, are marked at the top left of the reference spectrum in their unshifted positions. These two lines are shifted toward the red (to the right in the photographs) by increasing amounts for more distant galaxies.

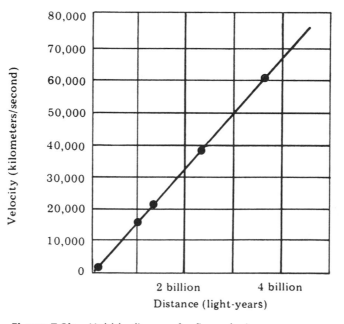

Figure 7.3b Hubble diagram for five galaxies.

Use Figure 7.3a. Plot a rough graph where each point represents the velocity of recession and distance of one of these galaxies. What do you observe when you draw a smooth curve through the five points? Explain. _____

_____

_____

Answer: The points all lie near a straight line (Figure 7.3b). That means there is a linear relationship between velocity of recession and distance from us for these galaxies.

7.4 VELOCITY-DISTANCE RELATION

U.S. astronomer Edwin Hubble, who spent most of his life studying galaxies, examined the relationship between the velocity of recession and the distance away for many galaxies. He discovered that the linear relationship you just found is true in general: The farther away a galaxy is, the faster it is receding.

The **Hubble law** (1929) says that a galaxy's velocity of recession, v, is directly proportional to its distance from us, d. The formula is:

$$v = Hd \qquad \text{where H is called the \textbf{Hubble constant.}}$$

The Hubble constant is very important. It gives the rate at which the galaxies are receding, or the rate at which the universe is expanding. It is also used in the Hubble law to figure the distance to galaxies from their measured redshifts.

To determine H accurately is difficult because of uncertainties in the extragalactic distance scale. The stated value is often updated and is roughly from 50 to 100 km/sec/Mpc (15 to 30 km per second per million light-years).

Some quasars have the largest redshifts ever observed (Section 6.20). If this phenomenon is a Doppler effect and these quasars are moving away faster than any known galaxies, what can you say about their distance? Explain. _____

_____

_____

Answer: These quasars are the most distant objects we can observe. The Hubble law says the most distant objects are those that are moving away the fastest.

7.5 PREMISE

The basic assumption we make in attempting to understand the universe is called the **cosmological principle**.

The cosmological principle states that on a sufficiently large scale the universe is **homogeneous** and **isotropic**. At any given time, the distribution of matter is the same everywhere in space, and the universe looks the same in all directions.

Our neighborhood in space is not particularly special. The laws of physics are universal. Any observer anywhere in the universe would see at any given time just about the same things we do on a large scale.

The cosmological principle is important because it lets us assume that the small portion of space we can see is truly representative of all the rest of the universe that we cannot see. It allows us to formulate a theory that explains the entire universe, including those parts we cannot observe.

Our observations show that distant galaxies are racing away from us. Does that mean that our Milky Way Galaxy is the center of the entire universe? Explain.

_____

_____

Answer: No. The cosmological principle says that if you went to any other galaxy and looked out into space, you would still see approximately equal numbers of galaxies located in all directions in space, racing away from you.

You can do a simple activity to illustrate this principle (Figure 7.4). Obtain a balloon and let its skin represent three-dimensional space. Mark dots randomly on the balloon to depict galaxies. Label one dot *MW* to represent our Galaxy. Blow up the balloon so it curves into a "fourth dimension." Observe how all the dots (galaxies) are pulled increasingly farther apart (recede from each other) by the stretching of the balloon's skin. (The "fourth dimension" is time. The volume inside the balloon represents past events, and that outside represents the future.)

Figure 7.4   An inflating balloon provides a conceptual model of the expanding universe.

7.6 STANDARD BIG BANG THEORY

The **Big Bang** theory says that the universe violently exploded into being in an event called the Big Bang and it has been evolving ever since. The beginning occurred 10 billion to 20 billion years ago.

All of the matter and radiation of our present universe were initially packed together in the **primeval fireball**, an extremely hot, dense state from which the universe rapidly expanded. The Big Bang was the start of that time and space we can know about.

The matter and radiation of that early inferno rapidly expanded and cooled. In a few seconds, protons (hydrogen nuclei), neutrons, and electrons were formed. Within minutes, the first deuterium (heavy hydrogen) and helium nuclei, and traces of a few light elements, were created.

Several million years later, matter and radiation were decoupled. Galaxies and stars began to form. The universe has continued to expand in space-time and the galaxies have continued moving away from each other ever since.

Today we observe that the universe is still expanding. Stars are still forming inside galaxies, using the original hydrogen from the Big Bang. The observed material of the universe is approximately 74 percent hydrogen and 24 percent helium, with traces of other light elements, such as deuterium and lithium, as predicted.

Most astronomers accept the Big Bang theory's description of the past and present stages of the universe. Predictions vary about the future, when the original hydrogen will finally be used up in stars and they all stop shining. The ultimate fate of the universe will be determined by the competition between the outward expansion and the inward pull of gravity.

The **open universe** model says that universe will continue to expand indefinitely. Then the universe, which began with a fiery Big Bang, will fade into darkness with a cold "whimper."

Refer to Figure 7.5. Identify and briefly describe the stages of an open universe according to the Big Bang theory. (a) _____ ; (b) _____ ; (c) _____ _____ ; (d) _____ .

Answer: (a) Big Bang explosion took place. (b) Galaxies formed. (c) Galaxies are still receding; the universe is expanding. (d) Original hydrogen will be used up; the resulting cold, black universe will continue expanding indefinitely.

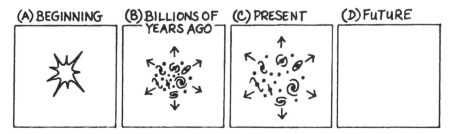

Figure 7.5 Stages of the open universe (Big Bang theory).

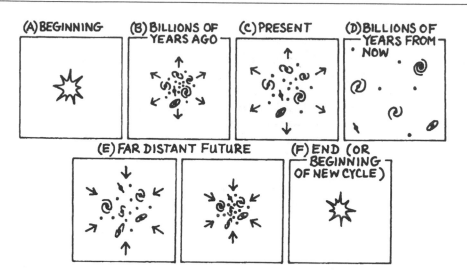

Figure 7.6   Stages of the closed (or oscillating) universe (Big Bang theory).

7.7 BIG CRUNCH

The **closed universe** model says that our universe, which began with the Big Bang, will not expand forever. Gravity will halt the expansion and force a collapse.

If the universe is closed, we happen today to be in the observed expanding phase. In the future our expanding universe will slow down, come to a complete stop, and then begin to contract. As it contracts, galaxies will fall back inward toward one another until all matter is once again crunched into an extremely hot, dense state.

The **oscillating universe** model variation says that after the Big Crunch, another Big Bang will occur. Then a new expanding universe will be born out of the same matter. The universe oscillates forever.

Refer to Figure 7.6. Briefly describe the stages of a closed universe according to the Big Bang theory. (a) _____ ;

(b) _____ ;

(c) _____ ;

(d) _____ ;

(e) _____ ;

(f) _____ .

Answer: (a) Big Bang occurred. (b) Galaxies formed and continued to recede. (c) We live in an expanding universe; galaxies are racing away from one another today. (d) Galaxies will stop. (e) The universe will contract; galaxies will fall back inward. (f) Matter will be crunched together again.

7.8 STEADY STATE THEORY

The **Steady State theory** was a rival of the Big Bang theory a few decades ago. It said that the universe does not evolve or change in time. There was no beginning in the past and there will be no end in the future. Past, present, future—the universe is the same forever.

This theory assumes the **perfect cosmological principle**, which says that the universe is the same everywhere on the large scale, at all times. It maintains the same average density of matter forever.

In order to explain the observation that the universe is expanding, the Steady State model says that new hydrogen is created continuously in empty space at a rate just sufficient to replace matter carried away by receding galaxies. The theory does not explain where the new hydrogen comes from.

Most astronomers reject the Steady State theory because it contradicts observations. Creation of unexplained new mass, a form of energy, violates a natural law—the conservation of energy law—which states that the total energy in an isolated system always remains the same. Energy cannot be created or destroyed, although transformations may occur within the system.

To its proponents, the Steady State theory had philosophical appeal. It defined a universe that always existed in the past and will always exist in the future. The hypothesis that heavy elements had to be made in exploding stars, advanced to explain how they got here without a Big Bang, survives intact in evolutionary models.

Refer to Figure 7.7. Briefly describe the universe according to the Steady State theory. (a) _____

_____

(b) _____

_____

Answer: (a) Galaxies are receding, the universe is expanding, new matter is being created, new galaxies are being formed. (b) The same pattern will occur. The universe maintains the same average density forever.

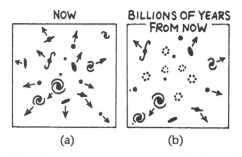

Figure 7.7   Stages of the universe (Steady State theory).

7.9 OBSERVATIONAL TESTS

Astronomers test a cosmological model by seeing whether it agrees with all the observational data we have about the universe.

The most direct way to check how the universe is evolving is to compare the way it looks today with the way it looked billions of years ago. Since we cannot actually make observations over billions of years as the universe ages, astronomers instead look at galaxies that are at different distances away from us.

Although the idea—to look back in time you study current images of distant galaxies—is simple, it is very hard to carry out in practice. Technology is not sufficiently developed to allow detailed imaging of very distant objects.

Consequently, all data that might be used to check cosmological models are full of uncertainties. No sufficiently precise data are yet available to confirm that any one of the models is entirely correct.

How can astronomers find out what the universe was like (a) 2 million years ago? (b) 3 billion years ago? Explain. _____

_____

_____

_____

_____

Answer: The most direct way would be to examine images of galaxies such as (a) Andromeda, which is about 2.2 million light-years away from Earth, and (b) Hydra, which is about 3 billion light-years away. It takes light one year to travel a distance of 1 light-year. The light we now receive left Andromeda 2.2 million years ago or Hydra 3 billion years ago; it tells us now what the universe was like then.

7.10 INCONSTANT HUBBLE CONSTANT

Cosmologists compare the value of the Hubble constant now and billions of years ago to verify their predictions.

Evidently the Hubble constant is not constant over time. It is decreasing, which means that the universe is slowing down in its expansion.

The Big Bang theory predicts a deceleration. The Hubble constant must be smaller now than it was billions of years ago because the components of the universe pull on one another due to gravitational attraction.

The Hubble constant decreases faster in the closed universe model than in the open universe model. If the universe is closed, it is slowing down at such a rate that it will actually come to a complete stop.

The **flat universe** model says that our universe will neither expand forever nor collapse. Gravity will force the deceleration and the velocity of expansion to become zero at the same time.

Results for the Hubble constant are too uncertain to be conclusive because no one can measure the distances to clusters of galaxies with high precision yet.

Why is it so important to measure the value of the Hubble constant very accurately? _____

_____

Answer: An accurate value of the Hubble constant would be strong evidence in favor of one of the cosmological models described. The Hubble constant is used to figure the age and size of the universe.

7.11 MATTER AND ENERGY

Observations of the density of matter (and energy) in the universe add vital clues for deciding among alternative cosmological models.

The **critical density** is the minimum average density of matter required for the force of gravity to stop the expansion of the universe without reversing it. The calculated value, which depends on the value of the Hubble constant (uncertain), is roughly 10^{-29} grams per cubic centimeter or a few hydrogen atoms per cubic meter.

The abundance of deuterium in space today sets a limit to the maximum possible amount of ordinary matter in our universe. Essentially all existing deuterium is presumed to have been created at the Big Bang so it is tightly linked to the initial density of matter. Observations indicate that the Big Bang created only 0.1 of the ordinary matter, and resulting gravitational force, that is necessary to eventually halt the observed expansion.

The average density of all matter presently observed by its radiation or gravitational effect is less than the critical value. The corresponding force of gravity cannot ever stop the expansion. The universe appears to be open.

Proponents of a closed universe have been searching for **missing mass**, matter as yet undetected, which would have to exist for a higher-than-critical average density of matter. They speculate that unseen dark matter, if it exists, could be in an exotic form such as massive neutrinos, black holes, brown dwarfs, or **WIMPs**, weakly interacting massive particles.

What would be the cosmological significance of new discoveries of so-far-unobserved mass and energy in the universe? _____

_____

_____

Answer: The density of matter in the universe would be greater than the present observed value. It might be sufficient to stop the expansion of the universe, or to reverse it. The universe may be flat or closed, although it appears to be open today.

7.12 UNIVERSAL RADIATION

The Big Bang theory predicts that the universe should still be filled with **cosmic background radiation,** a remnant of radiation left over from the original Big Bang.

The primeval fireball would have sent strong shortwave radiation (corresponding to a temperature up to trillions of degrees) in all directions into space like exploding gigantic atomic bombs. In time, that radiation would spread out, cool, and fill the expanding universe uniformly. By now it would strike Earth as microwave (short radio) radiation, corresponding to a temperature only a few degrees above absolute zero.

In 1965 U.S. physicists Arno Penzias and Robert Wilson detected microwave radiation coming equally from all directions in the sky, day and night, all year. The radiation is like that radiated by a blackbody at a temperature of 2.7 K and remarkably uniform everywhere.

Apparently astronomers have detected the fireball radiation that was produced by the Big Bang.

What does the discovery of the cosmic background radiation mean for the steady state theory? _____

_____

Answer: It invalidates the steady state model, since the model cannot explain the existence of this radiation.

7.13 BIG BANG MODEL SUCCESSES

Summarize the observations that the Big Bang model successfully explains.

_____

_____

_____

Answer: Your answer should include the following observations: (1) redshifts of distant galaxies; (2) cosmic background radiation; (3) abundances of hydrogen and helium.

7.14 BIG BANG QUESTIONS

The standard Big Bang model fails to explain how an explosive beginning resulted in both the homogeneity of the cosmic background radiation and the large-scale structure of the observable universe.

If the initial distribution of energy and mass was smooth, gravity alone could not clump ordinary matter into the observed large clusters and superclusters of galaxies in the calculated lifetime of the universe. Probably there were some initial anisotropies and inhomogeneities.

In 1981, U.S. physicist Alan Guth proposed **inflation**, a brief phase of

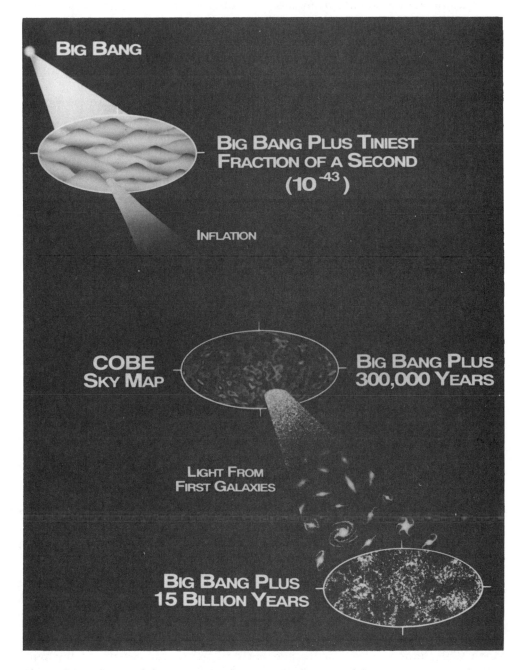

Figure 7.8 History of the universe (inflationary Big Bang model).

incredibly rapid expansion shortly after the Big Bang, to account for the present vast extent of the universe and its uniformity. The **flatness problem**, how to explain why the earliest density of the universe must have been extraordinarily close to the critical density, is resolved by adding the **inflationary universe** model to the Big Bang theory.

Recent measurements from the U.S. robot satellite Cosmic Background Explorer (COBE) show slight temperature variations in the average sky temperature of 2.7 K. These variations represent broad ripples of wispy matter which could have grown into galaxies, clusters of galaxies, and the great voids in space today (Figure 7. 8).

The preferred **Einstein-de Sitter** model of inflationary cosmology calls for a homogeneous, flat universe. Proponents hypothesize that an enormous amount of matter in the universe is dark, exotic and escaping detection.

Future observations will test how much and what kinds of dark matter actually exist.

Few astronomers want to abandon Big Bang theory entirely. What are two major concerns today? (1) _____

_____

(2) _____

_____

Answer: (1) A model of how the universe evolved after the first moments of the Big Bang that is consistent with the observed large-scale structure in the universe. (2) Unambiguous detection of dark matter, in a familiar or exotic form.

7.15 AGE

Estimates of the age of the universe have tended to grow from a biblical few thousand years to millions and then billions of years.

Standard estimates of the age of the universe are based on the value of the Hubble constant. The **Hubble time**, the age of the universe since the time of the Big Bang, is equal to 1/H. The calculated age depends greatly on the value of the (still imprecise) Hubble constant, with some correction for the slowing down of the universe in the past. This method puts the time of the Big Bang at from 10 billion to 20 billion years ago.

Radioactive dating of rocks and meteorites yields roughly similar ages. U.S. astronomer David N. Schramm figured the age of the universe to be 20 billion years by calculating how much radioactive rhenium 187 had decayed since that element was first formed early in the history of the Milky Way Galaxy. Still another age is derived from that of the oldest stars. By this method, the universe is from 13 billion to 18 billion years old.

List three methods of estimating the age of the universe.(1) _____

_____

(2) _____

(3) _____

Answer: (1) Measuring the Hubble constant and Hubble time of 1/H; (2) radioactive dating of rocks and meteorites; (3) estimating based on the age of the oldest stars observed.

7.16 SIZE

Estimates of the radius of the universe also depend very much on the value of the Hubble constant. The **Hubble radius**, the distance to the edge of the observable universe, is equal to the speed of light divided by the Hubble constant, c/H. These estimates put the radius of the universe between 12 billion and 16 billion light-years (Figure 7.9).

The eternal questions of human beings—how did the world begin, and will it end?—cannot be answered by science today. Review the range of estimates we have about the universe by completing the following chart:

The Observable Universe Today

(a) Rate of recession of distant galaxies (Hubble constant)_____

_____

(b) Approximate radius _____

(c) Age in approximately present form_____

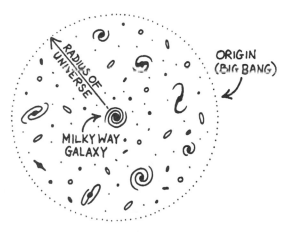

Figure 7.9   The Hubble radius.

Answer:

(a) The rate of recession (Hubble constant) is roughly from 50 to 100 km/sec/Mpc (15 to 30 km per second per million light-years).

(b) From 12 billion to 16 billion light-years.

(c) From about 10 billion to 20 billion years old.

SELF-TEST

This self-test is designed to show you whether or not you have mastered the material in Chapter 7. Answer each question to the best of your ability. Correct answers and review instructions are given at the end of the test.

1. Define cosmology._____

   _____

2. How do cosmological models differ from religious explanations of the universe?_____

   _____

   _____

3. Describe the evidence that the universe is expanding. _____

   _____

   _____

   _____

   _____

4. State the Hubble law. _____

   _____

   _____

   _____

   _____

5. Why is the Hubble constant so important in cosmology? _____

   _____

   _____

   _____

6. State the basic assumption of all cosmological models. _____

   _____

   _____

7. Match one or more of the three main cosmological models (Big Bang theory) with each statement:

____ (a) 10 billion to 20 billion years ago the universe exploded into being from an extremely dense hot state.

(1) Open.

(2) Closed.

____ (b) An enormous amount of matter in the universe is dark, exotic and escaping detection.

(3) Flat.

____ (c) Galaxies are moving away from each other with speeds that increase with distance.

____ (d) In the future, the universe will expand indefinitely.

____ (e) In the future, the universe will stop expanding and then contract.

8. List two basic observations that can help decide between an open and closed universe.   (1) _____ ;
(2) _____

9. What is the cosmological significance of cosmic background radiation?

_____

_____

_____

10. State the main contribution of an inflationary universe model to Big Bang theory. _____

_____

_____

_____

11. Give the approximate (a) Hubble age of the universe _____
_____ ; (b) Hubble radius _____

ANSWERS

Compare your answers to the questions on the self-test with the answers given below. If all of your answers are correct, you are ready to go on to the next chapter. If you missed any questions, review the sections indicated in parentheses following the answer. If you missed several questions, you should probably reread the entire chapter carefully.

1. Cosmology is the branch of science concerned with the origin, present structure, evolution, and final destiny of the universe.   (Section 7.1)

2. Cosmological models do not give a supernatural cause or meaning to physical events, but try to explain these events using only the laws of nature and mathematics.   (Section 7.1)

3. Light from distant galaxies is shifted in wavelength toward the red end of the spectrum, a phenomenon called the redshift. The farther away a galaxy is, the greater its redshift. The most distant galaxies are receding uniformly from us and from each other.   (Section 7.2)

4. The Hubble law says that a galaxy's velocity of recession (v) is directly proportional to its distance away from us (d). The Hubble law can be written algebraically as

   $v = Hd$ where *H* is called the Hubble constant.   (Section 7.4)

5. The Hubble constant is very important because it gives the rate at which the galaxies are receding, or the rate at which the universe is expanding. It is a basis for estimating the size and age of the universe.   (Sections 7.4, 7.10, 7.15, 7.16)

6. The cosmological principle states that on a sufficiently large scale, the universe is homogeneous and isotropic at any given time.   (Section 7.5)

7. (a) 1, 2, and 3;   (b) 2 and 3;   (c) 1, 2, and 3;   (d) 1;   (e) 2.   (Sections 7.2, 7.6, 7.7, 7.10, 7.11, 7.14)

8. (1) Rate of change of the Hubble constant with time;   (2) density of matter in the universe.   (Sections 7.10, 7.11)

9. The microwave radiation striking Earth from all directions in space provides strong evidence for the big bang model. It appears to be the redshifted remnant of the radiation created in the Big Bang.   (Section 7.12)

10. A brief phase of incredibly rapid expansion shortly after the Big Bang could explain how an explosive beginning could result in both the homogeneity of the cosmic background radiation and the large-scale structure of the observable universe.   (Section 7.14)

11. (a) 10 billion to 20 billion years;   (b) from 12 billion to 16 billion light-years.   (Sections 7.15, 7.16)

8

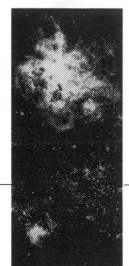

EXPLORING THE SOLAR SYSTEM

Finally we shall place the Sun himself at the center of the Universe.

Nicolaus Copernicus
De Revolutionibus Orbium Coelestium (1543)

Objectives

☆ List the members of the solar system.
☆ State the essential difference between a planet and a star.
☆ Describe evidence supporting the nebular theory of the formation of the solar system.
☆ Explain the phases of the Moon.
☆ Outline the development of our understanding of the solar system, including the contributions of Ptolemy, Copernicus, Galileo, Tycho Brahe, Kepler, and Newton.
☆ State and apply the laws governing the motions of bodies under gravity.
☆ Explain the apparent motions of the planets, including retrograde motion.
☆ Explain why the Moon's sidereal month and synodic month differ.
☆ Differentiate between the revolution and rotation of celestial bodies.
☆ Explain the motions of Earth-orbiting satellites and interplanetary spacecraft.
☆ Compare and contrast the general properties of the nine large planets and their moons.
☆ Describe the asteroids (minor planets).

8.1 INVENTORY

Our **solar system** includes the Sun and all objects gravitationally bound to it—nine planets with their moons, asteroids (also called minor planets), comets, and interplanetary dust and gas.

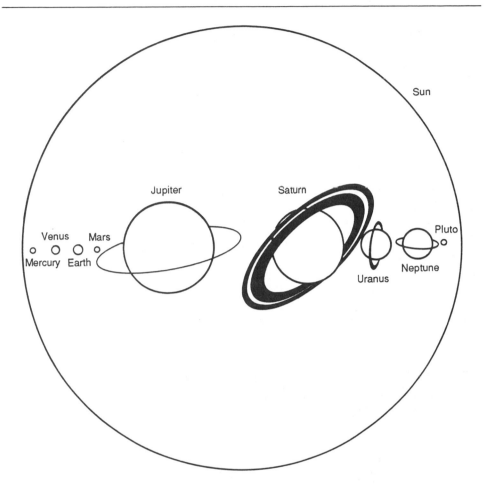

Figure 8.1 Relative sizes of the planets. (Not to scale with Sun.)

Planets are bodies that orbit stars directly and **moons** are bodies that orbit planets. Planets and moons are much less massive and colder than stars. Stars generate their own light, but planets and moons are not massive enough for nuclear fusion reactions to ignite. Planets and moons shine by reflecting starlight.

The planets of our solar system range in mass from lightest Pluto to heaviest Jupiter 318 times Earth's mass. All of the planets together have only 0.001 the mass of the Sun.

In size, the planets range from smallest Pluto to largest Jupiter, whose diameter is 0.1 of the Sun's (Figure 8.1).

What is the essential difference between a planet and a star? _____

_____

Answer: A planet is much less massive and colder than a star. It shines by reflecting light from a star. (A star generates its own light.)

8.2 ORIGIN OF THE PLANETS

The **solar nebular model** says that the solar system formed out of an eastward rotating interstellar cloud about 5 billion years ago (Section 4.3). The nebula contracted into the proto-Sun surrounded by a spinning disk where the planets formed (Figure 4.4). The new Sun blew away most residual gas and dust.

The nebular theory is supported by the properties of the solar system today.

All of the planets **revolve**, or travel around, the Sun in the same direction from west to east, or counterclockwise as seen from above. This movement is called **direct motion** (Figure 8.2). The planets rotate as they revolve. Their rotation (except for Venus and Uranus) is also direct.

The mean plane of Earth's orbit around the Sun is called the **ecliptic**. The orbits of all of the planets are in nearly the same plane, like the lanes on a running track. Pluto's **inclination**, the angle between its orbital plane and the ecliptic, is exceptional (Table 8.2).

The planets whose orbits are closer to the Sun than Earth's are called **inferior**, while those whose orbits are outside Earth's are called **superior**.

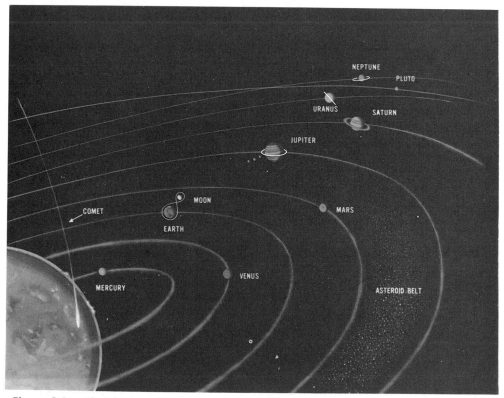

Figure 8.2. The order of the planetary orbits (not to scale).

Refer to Figure 8.2. List the (a) inferior and (b) superior planets.

(a) _____

(b) _____

_____

Answer: (a) Mercury, Venus; (b) Mars, Jupiter, Saturn, Uranus, Neptune, Pluto.

8.3 DAY NAMES

Five planets—Mercury, Venus, Mars, Jupiter, and Saturn—look like very bright stars in the sky. The Sun, Moon, and these five bright planets were known to the ancients. Each was thought to rule one day of the week, which was given its name (in Latin).

We take the names for our days of the week from the Anglo-Saxons, who substituted the names of equivalent gods and goddesses for the Roman ones. French and Spanish names are adapted directly from the Latin.

Refer to Table 8.1. (a) Which of our days of the week is closest to the original Latin god's name? _____ (b) Which days of the week carry the names of Anglo-Saxon gods? _____

Answer: (a) Saturday; (b) Tuesday, Wednesday, Thursday, Friday.

TABLE 8.1 Days of the Week

Day	Ruling Planet	Anglo-Saxon Equivalent	Latin	French	Spanish
Sunday	Sun	—	*Dies Solis*	*Dimanche*	*Domingo*
Monday	Moon	—	*Dies Lunae*	*Lundi*	*Lunes*
Tuesday	Mars	Tiw	*Dies Martis*	*Mardi*	*Martes*
Wednesday	Mercury	Woden	*Dies Mercurii*	*Mercredi*	*Miercoles*
Thursday	Jupiter (Jove)	Thor	*Dies Jovis*	*Jeudi*	*Jueves*
Friday	Venus	Frigg	*Dies Veneris*	*Vendredi*	*Viernes*
Saturday	Saturn	Seterne	*Dies Saturni*	*Samedi*	*Sabado*

8.4 MOON PHASES

The Moon is Earth's only natural satellite, orbiting our planet as we travel around the Sun. It shines in the sky by reflecting sunlight.

The Moon's appearance changes regularly every month. Half of the Moon is always lighted by the Sun, but the bright shape we see from Earth, called its **phase,** changes as the Moon travels around our planet. The recurring cycle of apparent shapes is called the **phases of the Moon.**

Refer to Figure 8.3. The **new Moon** is dark. It is not seen in the sky because the Moon's dark side is facing Earth. The **waxing** (growing bigger), **crescent** Moon follows a few days later. You often see its disk faintly lighted by sunlight reflected from Earth, called **earthshine.**

About 7 days after the new Moon, when the Moon has traveled 1/4 of its way around Earth, it rises around noon, and we observe **first quarter** shine. The **waxing gibbous** Moon follows, with more than half of the Moon's bright disk shining toward Earth.

When the Moon is about two weeks into its cycle, the **full Moon** lights the sky all night with its whole, bright disk. Full Moon occurs 12.37 times a year.

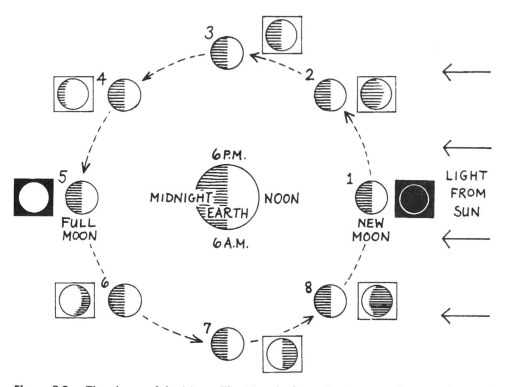

Figure 8.3. The phases of the Moon. The Moon's elongation is counted eastward around the sky. Elongations of 0°, 90°, 180°, and 270° correspond to new, first quarter, full, and last quarter Moon, respectively.

So months having two full Moons occur every 2.72 years on the average. The second full Moon within a particular month is called **blue Moon.** ("Once in a blue moon" means "very seldom.") Approximately once every 19 years, a year will have two months with two full Moons because February will not have a full Moon.

The visible part of the Moon's bright disk **wanes,** or decreases in size, as the Moon completes its trip around Earth during the last two weeks of its cycle.

The mean time required for the Moon's phases to repeat, called a **synodic month** or **lunation,** is 29.5 days. In this context, the age of the Moon is the time since new Moon.

Identify the phase of the Moon corresponding to the indicated position in its orbit for each of the following phases, as labeled in figure 8.3: (a) waxing crescent _____ ; (b) first quarter _____ ; (c) waxing gibbous _____ ; (d) waning gibbous _____ ; (e) third quarter _____ ; (f) waning crescent _____.

Answer: (a) 2; (b) 3; (c) 4; (d) 6; (e) 7; (f) 8.

Observe the Moon daily for a month, if possible. Keep a record of its bright shape, its position relative to the Sun, and its time of rising and setting. Use the information in this chapter to explain the changes you observe.

8.5 PLANET WATCHING

Planet positions are not marked on star maps. Stars keep their same relative positions in the sky for decades, but planets do not. The word "planet" comes from the Greek for "wanderer." Bright planets move across the celestial sphere near the ecliptic.

Venus and Mercury appear to move forward and backward on either side of the Sun in Earth's sky. Maximum **elongation,** or distance east or west of the Sun, is 48° for Venus and 28° for Mercury.

Mars, Jupiter, and Saturn wander generally eastward through the constellations of the zodiac. At times planets seem to reverse and move westward, called **retrograde motion**, before resuming direct motion. The apparent backward swing plus resumed forward motion is called a **retrograde loop** (Figure 8.4).

You can look up the exact locations of the planets on any given night in astronomical publications, computer software, and almanacs (see "Useful Resources").

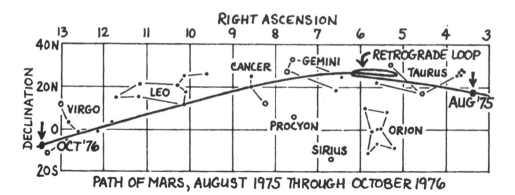

RIGHT ASCENSION

PATH OF MARS, AUGUST 1975 THROUGH OCTOBER 1976

Figure 8.4. Successive observed positions of Mars among the zodiac constellations during the time of the historic U.S. Viking flights to the red planet.

Suggest a way of finding a particular planet such as Jupiter in the sky tonight.

_____

_____

_____

_____

_____

Answer: Find out which zodiac constellation Jupiter is in today by using an astronomical publication, computer software, or an almanac. Locate that constellation on your star maps. For example, suppose you find Jupiter is in Taurus. If Taurus is in the sky tonight, you can easily spot Jupiter. It will be the brilliant "star" that does not belong to the constellation (Figure 8.5).

★ Keep a record of the position of Venus and Mars for several months. Observe them in the sky, if possible. Use the information in this chapter to explain the motions you observe.

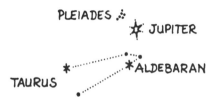

Figure 8.5.   Jupiter in the constellation Taurus.

8.6 BRIEF HISTORY

The search for a simple explanation of the planets' observed motions in the sky changed humanity's view of the world.

In *Almagest*, written about A.D.. 150, the Alexandrian astronomer Ptolemy (Claudius Ptolemaeus) described the ancient **geocentric**, or Earth-centered, **model** of the universe. Circles were considered "perfect" shapes. The Sun, Moon, and planets were supposed to move on small circles called **epicycles**, whose centers moved around Earth on larger circles called **deferents**.

For more than 14 centuries, this **Ptolemaic system** was accepted as the basis for astronomical work. It described with considerable accuracy the observed positions and motions of the heavenly bodies known at that time. It also expressed the view of the world that people had from observing the sky. With minor modifications, this Earth-centered theory became part of the dogma of the Roman Catholic Church of the Middle Ages.

Polish astronomer Nicolaus Copernicus (1473–1543) published his radical **heliocentric**, or Sun-centered, **model** the year he died. In the **Copernican system**, the planets, including Earth, circle around a stationary central Sun. According to the Copernican model, the apparent wandering motions of the

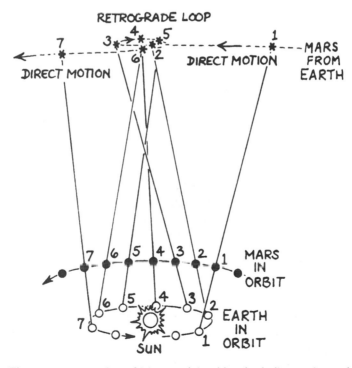

Figure 8.6. The apparent motion of Mars explained by the heliocentric model. Numbers mark positions at one-month intervals. (1–3): Mars appears to slow its direct motion as Earth overtakes it. (4): Mars appears to move backward as Earth passes it. (5–7): Mars resumes direct motion as Earth moves ahead of it.

planets result from a combination of the real orbital motions of both Earth and observed planets.

The apparent motion of Mars illustrated in Figure 8.6 is explained as follows. Mars never really moves backward in its orbit. Planets always go forward. The retrograde loop in the sky is caused by the relative motion of Earth and Mars. Our faster-moving Earth catches up to Mars and moves past it. So Mars, the outer planet, looks to us as if it is moving backward. It is like watching a slower car while you are overtaking it in a faster car.

What change in the philosophical view of Earth was required for people to accept the Copernican theory instead of the Ptolemaic? _____

Answer: Earth could no longer be considered the center of the entire universe and supremely important.

8.7 FIRST TELESCOPIC DATA

Italian scientist Galileo Galilei (1564–1642) was the first person to use a telescope for sky observations. His observations provided important data for the Copernican theory.

The sight of mountains, craters, and extensive dark areas on the Moon, as well as sunspots and their movements, convinced Galileo that the heavens were not perfect and unchanging. His discovery of four large moons orbiting Jupiter confirmed that Earth was not the center of all heavenly motions.

Galileo observed that bright Venus appears to change its shape and size regularly. The Ptolemaic system could not account for the phases of Venus. But the Copernican system had a simple explanation.

Refer to Figure 8.7. Venus and Mercury, the two inferior planets, show phases as they reflect sunlight to Earth from different places in their orbits around the Sun. An inferior planet looks most fully lit in its gibbous phase, near **superior conjunction**, the point on the far side of the Sun from Earth. It appears as a crescent and biggest near **inferior conjunction**, the point between Earth and Sun.

The Roman Catholic Church in 1616 banned books that supported Copernicanism. Galileo was allowed to continue research, provided he did not hold, teach, or defend heretical doctrines. But his "Dialogue Concerning the Two Chief World Systems," published in 1632, supported the Copernican system.

The next year, Galileo, almost 70 years old and after a brilliant career, was forced by the Inquisition to recant his astronomical findings and condemned to house arrest. Debates about the compatibility of religion and science raged among scientists, theologians, and historians for 350 years. The Church slowly acted to rectify the infamous verdict and finally vindicated Galileo in 1992.

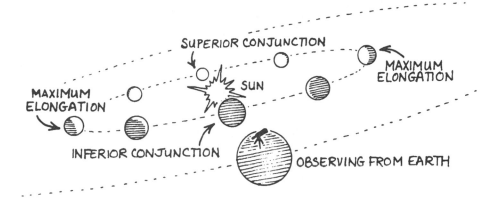

Figure 8.7. The phases of Venus as viewed through a telescope from Earth.

8.8 LAWS OF PLANETARY MOTION

German astronomer Johannes Kepler (1571–1630) deduced a simple, precise description of planetary motion. Kepler worked from records inherited from Tycho Brahe (1546–1601), a Danish astronomer who had recorded the positions of stars and planets with unprecedented accuracy for almost 20 years. The remains of Tycho's observatory and artifacts are on display on the now-Swedish island of Ven.

Kepler's laws of planetary motion greatly improved the accuracy of predictions of planet positions. The three laws state:

1. Each planet moves around the Sun in an orbit that is an ellipse with the Sun at one focus.

2. Each planet moves so that an imaginary line joining the Sun and the planet sweeps out equal areas in equal times. As illustrated in Figure 8.8, the planet goes from A to B and C to D in the same amount of time. In other words, planets move fastest when they are closest to the Sun (**perihelion**) and slowest when they are farthest away (**aphelion**).

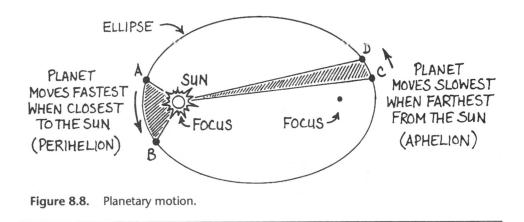

Figure 8.8. Planetary motion.

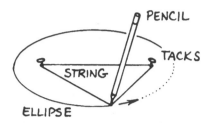

Figure 8.9. Drawing an ellipse.

3. The squares of the periods of time required for any two planets to complete a trip around the Sun have the same ratio as the cubes of their average distances from the Sun.

Kepler's third law can be used to find a planet's average distance, d, from the Sun compared to Earth's average distance of 1 AU (Section 4.2). The planet's orbital period, p, in years is found from observations. Then Kepler's third law is written $d^3 = p^2$.

For example, Jupiter's orbital period is 11.86 years. So Jupiter's average distance, d, from the Sun is found from $d^3 = (11.86)^2 \cong 141$. Solving for d = $\sqrt[3]{141}$, we find d = 5.2 AU.

How far would a planet be from the Sun if its orbital period were observed to be 8 years? Explain. _____

Answer: 4 AU, or 4 times Earth's average distance. According to Kepler's third law, $d^3 = p^2$. So $d^3 = (8)^2 = 64$. And d = $\sqrt[3]{64}$ = 4.

Note: An **ellipse** is a closed curve for which the sum of the distances from any point on the curve to two fixed points, or **foci**, inside is a constant. The **semi-major axis**, half the maximum dimension (major axis), defines size. **Eccentricity**, or distance between foci divided by major axis, measures how much the ellipse deviates from being a circle.

💡 To draw an ellipse, put two tacks for foci in a board. Tie a string around them. Trace the ellipse by keeping the string taut with your pencil point (Figure 8.9).

8.9 MOTION AND GRAVITY

Kepler's laws explain how the planets are observed to move. English physicist and mathematician Sir Isaac Newton (1642–1727) formulated laws that explain why the planets move as they do. His book *The Mathematical Principles of Natural Philosophy* was published in 1687.

Newton's laws of motion state:

1. A body stays in a state of rest or uniform motion unless an outside force acts on it.

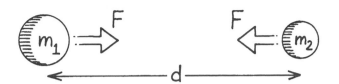

Figure 8.10. Newton's law of gravity.

2. The net force, F, on a body is equal to its mass, m, multiplied by its acceleration, a. The formula is: F = ma.

3. Whenever one body exerts a force on a second body, the second body exerts an equal and opposite force on the first body.

Newton's **law of gravity** states: Any two objects of masses m_1 and m_2 separated from each other by a distance, d, attract each other with a force, F, called gravity, that is directly proportional to the product of their masses and inversely proportional to the square of their distance away from each other (Figure 8.10). The formula is:

$$F = \frac{G\, m_1 m_2}{d^2} \quad \text{where G = gravitational constant (Appendix 2).}$$

A force of attraction is necessary to keep the planets moving in their curved paths around the Sun. Without this force they would move straight away into space. The required force is provided by the Sun's gravity, which continuously pulls the planets in toward the Sun.

The combination of their forward motion and their motion in toward the Sun under gravity keeps the planet traveling in its orbit around the Sun (Figure 8.11).

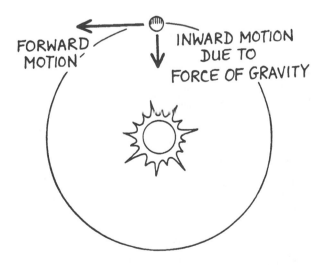

Figure 8.11. The two motions that keep a planet traveling in orbit.

Newton's genius was to realize that his law of gravity applies to falling objects on Earth, to the motion of the Moon and planets, and to all material bodies. He proposed that his law of gravitation and his three laws of motion were basic laws of physics. His laws are **universal**, true for all objects everywhere in the universe.

Newton generalized and mathematically derived Kepler's laws of planetary motion from basic principles. He invented and used in his work the branch of mathematics we call calculus.

Apply Newton's laws to explain why moons stay in orbit around their parent planets. _____

_____

_____

Answer: A combination of two motions keeps a moon in orbit around its parent planet— its forward motion and its inward motion caused by the pull of the planet's gravity.

8.10 MOON'S ORBITAL MOTION

The Moon's orbit is an ellipse with Earth at one focus. Its travels at an average speed of 1.02 km/sec (2295 miles per hour).

The time for the Moon to complete one trip around Earth with respect to the stars, about 27.3 days, is called a **sidereal month**.

The Moon's average angular diameter in the sky is about ½° (31'5″ of arc). The Moon looks larger at **perigee**, the point in its orbit closest to Earth, and smaller at **apogee**, the point in its orbit farthest from Earth.

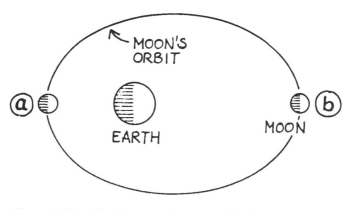

Figure 8.12.   The Moon's orbit around Earth.

Identify the apogee and perigee in Figure 8.12, and indicate where the Moon looks larger than average and where it looks smaller than average. (a) _____ _____ ; (b) _____

Answer: (a) Perigee—the Moon looks larger; (b) apogee—the Moon looks smaller.

8.11 SYZYGY

If you enjoy word games or crossword puzzles, you'll find **syzygy** a good word to know. It means three celestial bodies in a line, such as Sun-Moon-Earth.

Refer to Figure 8.13. Explain why a synodic month, or the month of the Moon's phases, is 2 days longer than a sidereal month. _____

_____

_____

_____

Answer: Start with new Moon (1). After 27.3 days, the Moon has traveled completely around Earth (2). But Earth and Moon have also moved together around the Sun during that time. Two more days must pass before the Moon, Earth, and Sun are lined up so that the Moon is new again (3).

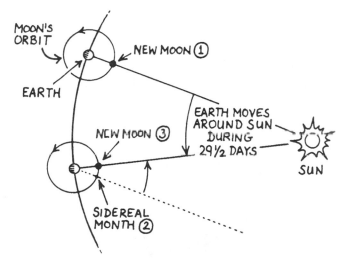

Figure 8.13.   The interval from one new Moon to the next is 2 days longer than a sidereal month.

8.12 SPACEFLIGHT

Spacecraft obey the same basic laws of physics that natural astronomical bodies do.

Any body in orbit around a larger parent body is called a **satellite**. Rockets launch artificial satellites into Earth-orbit with a forward velocity of at least 8 km/sec (17,300 miles per hour). The combination of their forward motion and their motion in toward Earth under Earth's gravity keeps the satellites in their orbits. Most are designed to burn up due to friction if they plunge back into the atmosphere. Piloted craft and selected payloads are made to survive re-entry and land safely.

Russian Sputnik 1, an 82-kg (180-pound) metal ball with a transmitter and batteries was the first artificial satellite. Rocketed to space on October 4, 1957, Sputnik 1 signaled the opening of the space age. Today hundreds of robot communications, weather, research, navigation, and military satellites of many nations are operating in Earth orbits.

Robot spacecraft are sent to explore the planets. These spacecraft are launched with a forward velocity into orbit around the Sun. Their motions are calculated using Newton's laws, just as planetary motions are.

All of the planets except Pluto have been observed close up by robot space-

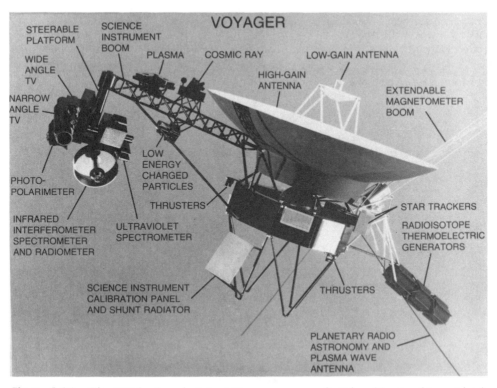

Figure 8.14. The 800-kg (nearly 1 ton) Voyager spacecraft. It has 11 sets of target body (point at object) and particles, fields, and waves sensors. Nuclear-electric generators power the spacecraft instruments, radio, and computers.

craft. These planet probes are packed with cameras, data sensors, and computers programmed to operate automatically far from direct human interaction. None has returned so far. They radio images and data back to Earth for analysis.

The most ambitious multitargeted space flight yet is the U.S. Project Voyager. Twin spacecraft, Voyager 1 and Voyager 2 (Figure 8.14), were launched in 1977 to take advantage of a planet lineup that occurs once in 176 years. The 12-year mission included encounters with Jupiter, Saturn, Uranus, and Neptune, their unique ring systems, and 48 of their moons.

Down-to-the-minute trajectory planning was done. **Gravity assist**, a technique of using a planet's gravitational field to change a spacecraft's velocity without consuming fuel, was used at each encounter to increase Voyager's speed and bend its flight path enough to convey it to the next destination.

Voyagers 1 and 2 returned a total of 118,000 images and data that have revolutionized the science of planetary astronomy. Now Voyager 1 is leaving the solar system, rising above the ecliptic plane at an angle of about 35° at a rate of about 520 million km (320 million miles) per year. Voyager 2 is speeding out below the ecliptic plane at an angle of about 48° at some 470 million km (290 million miles) per year.

Refer to Figure 8.15. About how many years after launch did Voyager 2 reach (a) Jupiter? _____ (b) Saturn? _____ (c) Uranus? _____ (d) Neptune? _____

Answer: (a) 2 years; (b) 4 years; (c) 8.5 years; (d) 12 years.

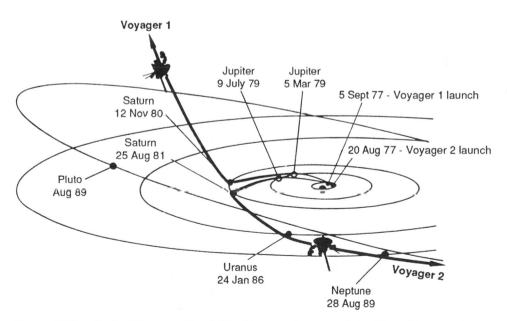

Figure 8.15. Project Voyager timetable. Voyager 2 was launched first. Voyager 1 was launched 16 days later on a faster, shorter trajectory. Encounter dates are marked where spacecraft and planet paths intersected.

TABLE 8.2 Properties of the Planets

	Mercury	Venus	Earth
Mean distance from Sun			
millions km	57.9	108.2	149.6
(millions of miles)	(36)	(68)	(93)
Astronomical Units, AU	0.39	0.72	1.00
Mean orbital velocity, km/sec	47.87	35.02	29.79
Period of revolution, sidereal	87.97 days	224.70 days	365.26 days
Synodic (days)	116	584	—
Rotation period, sidereal	58.6d	243.017d	23h 56m 4s
(*days*, *hours*, *minutes*, *seconds*)			
Inclination of orbit to ecliptic	7°00'	3°24'	0°00'
Eccentricity of orbit	0.206	0.007	0.017
Oblateness	0	0	0.0034
Equatorial diameter, km	4,878	12,104	12,756
(miles)	(3,030)	(7,520)	(7,930)
Mass (Earth = 1)	0.06	0.82	1.00
Density, t/m^3	5.43	5.24	5.52
Surface gravity (Earth = 1)	0.38	0.91	1.00
Confirmed satellites	0	0	1 moon

8.13 PLANET SURVEY

Comparative planetology, the study of one planet compared with others, helps us to better understand our own world as well as the rest of our solar system. The general properties of the nine planets are listed in Table 8.2.

Mercury, Venus, Earth, and Mars have similar physical and orbital characteristics. They are called **terrestrial**, or earthlike, **planets**. Jupiter, Saturn, Uranus, and Neptune are also similar to one another and are called **giant** or **Jovian** (meaning Jupiter-like), planets. Mysterious Pluto does not really fit into either group.

Examine Table 8.2. How do the terrestrial planets differ from the giant planets in (a) distance from the Sun? (b) size? (c) mass? (d) density?

Terrestrial Planets	**Giant Planets**
(a) _____	(a) _____
(b) _____	(b) _____
(c) _____	(c) _____
(d) _____	(d) _____

TABLE 8.2   Properties of the Planets *(Continued)*

Mars	Jupiter	Saturn	Uranus	Neptune	Pluto
227.9	778.3	1425.5	2877.1	4508.1	5955.0
(142)	(486)	(889)	(1792)	(2796)	(3693)
1.52	5.20	9.53	19.23	30.14	39.81
24.13	13.06	9.65	6.80	5.43	4.74
686.98 days	11.86 years	29.46 years	84.01 years	164.79 years	247.69 years
780	399	378	370	367	367
24h 37m 26s	9.842h	10.656h	17.232h	16.104h	6.39h
1°48′	1 °18′	2°30′	0°48′	1°48′	17°6′
0.093	0.048	0.053	0.047	0.006	0.255
0.0052	0.067	0.008	0.023	0.017	0?
6,787	142,980	120,540	51,120	49,530	2300
(4,220)	(88,850)	(74,900)	(31,770)	(30,780)	(1430)
0.11	317.89	95.18	14.54	17.15	0.002
3.94	1.33	0.69	1.27	1.64	2.0
0.38	2.54	1.08	0.91	1.19	0.06
2 moons	16 moons	18 moons	15 moons	8 moons	1 moon
	Rings	Rings	Rings	Rings	

Answer:

	Terrestrial Planets	**Giant Planets**
(a)	Near the Sun.	Far from the Sun.
(b)	Small diameter.	Large diameter.
(c)	Small mass.	Large mass.
(d)	High density.	Low density.

8.14 DAYS AND YEARS

The **period of revolution** is the length of time for a celestial body to go around its orbit once.

A planet's **sidereal revolution period** is measured relative to the stars. It is the length of the planet's year in terms of Earth time. A planet's **synodic revolution period** is the planet's orbital period as seen from Earth. It is equal to the time for a planet to return to a specific **aspect**, or certain position relative to the Sun as seen from Earth, such as conjunction.

A planet's synodic period differs from its sidereal period because Earth is itself moving in its orbit around the Sun.

The **period of rotation** is the length of time required for a celestial body to turn around on its axis once. A planet's **sidereal rotation period** is the length of one sidereal day on the planet (Section 1.23). A planet's **synodic rotation period** is the length of one solar day on the planet. It is the time interval between two successive meridian transits of the Sun as would be seen by an observer on that planet.

Earth's rotation was long used for timekeeping. But Earth's rotation is not exactly uniform. Its accuracy is good to about 0.001 second a day. Atomic clocks, which operate by measuring the resonant frequency of a given atom—cesium, hydrogen, or mercury—are accurate to a billionth of a second a day. The International Earth Rotation Service monitors the difference in the two time scales and adds one-second steps, called **leap seconds**, to the world's clocks as necessary.

Examine Table 8.2. (a) Which giant planet has the longest year, and how many Earth-years is it equal to? (b) Which planet has the longest sidereal day, and how many Earth-days is it equal to? _____

_____

_____

Answer: (a) Neptune. It is equal to 164.8 Earth-years. (b) Venus. It is equal to 243 Earth-days.

8.15 ASTEROIDS

Asteroids, or minor planets, are small, irregularly shaped rocky bodies that orbit the Sun. Most follow paths inside a region called the **asteroid belt**, which is between the orbits of Mars and Jupiter.

Through a telescope, asteroids (from the Greek for "starlike") look like stars. The largest asteroid was the first discovered by Sicilian astronomer Giuseppi Piazzi (1746–1826) in 1801. Numbered 1 and named Ceres, it is 1000 km (625 miles) across. Some 5000 asteroids have been catalogued since then, and up to 200 are added each year. Dactyl, orbiting 243 Ida, was the first natural satellite discovered, in 1994. Their total mass is apparently less than 0.0001 that of Earth. Billions more tiny ones, counting dust particles, probably exist.

The first closeup of an asteroid, 951 Gaspra, was taken by the U.S. robot Galileo in 1991 (Figure 8.16). Gaspra is a chunk of rock with craters scattered over a surface that is covered by rubble and soil. It measures about 12 × 20 × 11 km (8 × 12 × 7 miles).

Gaspra rotates in a counterclockwise direction in just over 7 hours. The amount of sunlight that other asteroids reflect to Earth is observed to vary and repeat after several hours, which indicates that they, too, have irregular shapes and are rotating (Figure 8.16).

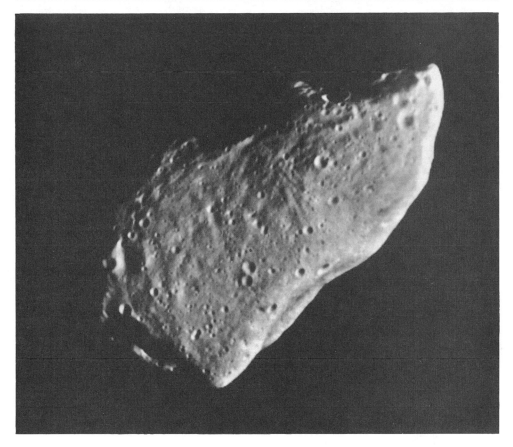

Figure 8.16. First closeup of asteroid, 951 Gaspra, from the U.S. robot Galileo when both were 55 million km from Earth. The smallest craters seen are some 300 m across.

Asteroids are classified into three main types using **spectrophotometry**, the accurate determination of magnitudes within specified wavelength regions. **C-type asteroids**, so called because they seem to be carbonaceous, are very dark and common in the outer asteroid belt. **S-type asteroids**, which seem to contain silicates mixed with metals, are moderately bright and are common in the inner belt. **M-type asteroids**, which appear to be metallic, are very bright.

The bright asteroids are probably clumps of mass that condensed from the original solar nebula, but never got big enough to form a large planet. The brightest, 4 Vesta, is 530 km (330 miles) in diameter. The fainter ones are probably fragments resulting from numerous collisions.

Some asteroids regularly come relatively near Earth. **Aten asteroids** have orbits inside Earth's. **Apollo asteroids** cross Earth's orbit and pass inside it to their perihelion. Apollo asteroids have come within a million kilometers of Earth. **Amor asteroids**, with orbits between 1 and 1.3 AU, stay beyond Earth's orbit.

When a nearby asteroid is first sighted, many people fear a disastrous collision. Astronomers judge asteroids that are larger than a kilometer in diameter as the greatest threat. Modern telescopes could probably spot such a large

TABLE 8.3   The Satellites of the Solar System

Planet	Satellite	Mean Distance from Planet[a] (km)	Diameter (km)	Revolution Period[b] (days)	Discovery
Earth	Moon	384,500	3,476	27.322	
Mars	Phobos	9,400	21	0.319	Hall, 1877
	Deimos	23,500	12	1.263	Hall, 1877
Jupiter	Metis	128,000	(40)	0.294	Synnott, 1980
	Adrastea	129,000	(25)	0.297	Jewett, Danielson, Synnott, 1979
	Amalthea	180,000	170	0.498	Bernard, 1892
	Thebe	222,000	(100)	0.674	Synnott, 1979
	Io	422,000	3,630	1.769	Galileo, 1610
	Europa	671,000	3,140	3.551	Galileo, 1610
	Ganymede	1,070,000	5,260	7.155	Galileo, 1610
	Callisto	1,885,000	4,800	16.689	Galileo, 1610
	Leda	11,110,000	(15)	240	Kowal, 1974
	Himalia	11,470,000	185	251	Perrine, 1904
	Lysithea	11,710,000	(35)	260	Nicholson, 1938
	Elara	11,740,000	75	260	Perrine, 1905
	Ananke	21,200,000	(30)	631	Nicholson, 1951
	Carme	22,350,000	(40)	692	Nicholson, 1938
	Pasiphae	23,330,000	(50)	735	Melotte, 1908
	Sinope	23,370,000	(35)	758	Nicholson, 1914
Saturn	Pan	134,000	(20)	0.577	Showatter, 1990
	Atlas	137,000	30	0.601	Terrile, 1980
	Prometheus	139,000	100	0.613	Collins, Carlson, 1980
	Pandora	142,000	90	0.628	Collins, Carlson, 1980
	Janus*	151,000	190	0.695	Dollfus, 1966
	Epimetheus*	151,000	120	0.695	Fountain, Larson, 1966
	Mimas	187,000	390	0.942	Herschel, 1789
	Enceladus	238,000	500	1.370	Herschel, 1789
	Tethys	295,000	1,060	1.888	Cassini, 1684
	Telesto	295,000	25	1.888	Smith, Larson, Reitsema, 1980
	Calypso	295,000	25	1.888	Pascu, Seidelmann Baum, Currie, 1980
	Dione	378,000	1,120	2.737	Cassini, 1684
	Helene	378,000	30	2.737	Laques, Lecacheux, 1980
	Rhea	526,000	1,530	4.517	Cassini, 1672
	Titan	1,221,000	5,150	15.945	Huygens, 1655
	Hyperion	1,481,000	255	21.276	Bond, Bond, Lassell, 1848

TABLE 8.3 The Satellites of the Solar System (Continued)

Planet	Satellite	Mean Distance from Planet[a] (km)	Diameter (km)	Revolution Period[b] (days)	Discovery
	Iapetus	3,561,000	1,460	79.331	Cassini, 1671
	Phoebe	12,960,000	220	550.46	Pickering, 1898
Uranus	Cordelia	49,800	25	0.33	Voyager 2, 1986
	Ophelia	53,800	30	0.38	Voyager 2, 1986
	Bianca	59,200	45	0.43	Voyager 2, 1986
	Cressida	61,800	65	0.46	Voyager 2, 1986
	Desdemona	62,600	60	0.48	Voyager 2, 1986
	Juliet	64,400	85	0.49	Voyager 2, 1986
	Portia	66,100	110	0.51	Voyager 2, 1986
	Rosalind	70,000	60	0.56	Voyager 2, 1986
	Belinda	75,300	68	0.62	Voyager 2, 1986
	Puck	86,000	155	0.76	Voyager 2, 1986
	Miranda	129,900	485	1.413	Kuiper, 1948
	Ariel	190,900	1,160	2.521	Lassell, 1851
	Umbriel	266,000	1,190	4.146	Lassell, 1851
	Titania	436,300	1,610	8.704	Herschel, 1787
	Oberon	583,400	1,550	13.463	Herschel, 1787
Neptune	Naiad	48,000	60	.30	Voyager 2, 1989
	Thalassa	50,000	80	.31	Voyager 2, 1989
	Despina	52,500	150	.33	Voyager 2, 1989
	Galatea	62,000	160	.43	Voyager 2, 1989
	Larissa	73,600	190	.55	Voyager 2, 1989
	Proteus	117,600	420	1.12	Voyager 2, 1989
	Triton	354,000	2,700	5.877	Lassell, 1846
	Nereid	5,510,000	340	365.21	Kuiper, 1949
Pluto	Charon	19,100	1,200	6.387	Christy, 1978

[a]At mean opposition distance.
[b]Sidereal period.
Parentheses indicate value is estimated.
*Coorbital satellites.

Source: Adapted and selected from National Aeronautics and Space Administration public information.

asteroid decades before it reached Earth so a catastrophe could be prevented by interception.

Analysis of reflected radar signals bounced off an asteroid recently confirmed the first metal asteroid with an orbit near Earth. Water, in the form of water of hydration, was first detected on 1 Ceres. Metallic asteroids near Earth could be mined in the twenty-first century to provide raw materials for space colonists and interplanetary expeditions.

Several enigmatic small, icy objects such as 2060 Chiron, 120 km (62 miles) in size, have been sighted in orbits around the Sun beyond the orbit of Neptune. Their origin is uncertain. They could be the first asteroids sighted in an outer zone. Chiron developed a diffuse gas envelope when it swung closest to the Sun in 1990, which suggests the existence of surface ice, so it could be a comet. These trans-Neptunian bodies could be the first direct evidence for the theorized Kuiper Belt, the source of short-period comets.

What are asteroids? _____

_____

Answer: Swarms of irregular, rocky bodies that orbit the Sun, mostly between the orbits of Mars and Jupiter.

8.16 MOON COMPARISONS

Giant planets are much more massive with stronger gravity than terrestrial planets. Hence they can more readily hold moons that formed or passed nearby.

Refer to Tables 8.2 and 8.3. (a) How many satellites do terrestrial planets have altogether? (b) How many confirmed satellites do the giant planets have? (c) List the moons of other planets which are larger than our Moon. (d) Which is the largest known moon in the solar system?

(a) _____

(b) _____

(c) _____

(d) _____

Answer: (a) Terrestrial planets have only three moons—Earth has one; Mars has two. (b) Giant planets have many moons and rings; 57 moons and four ring systems are confirmed. (c) Jupiter: Ganymede, Callisto, Io; Saturn: Titan. (d) Ganymede.

SELF-TEST

This self-test is designed to show you whether or not you have mastered the material in Chapter 8. Answer each question to the best of your ability. Correct answers and review instructions are given at the end of the test.

1. List the members of the solar system. _____

   _____

2. What is the essential difference between a star and a planet? _____

   _____

   _____

3. Give two facts that support the nebular theory of the formation of the solar

   system. _____

   _____

4. In midnorthern latitudes, which phase of the Moon would you see if the

   Moon were rising in the sky about (a) 6 P.M.? _____

   (b) noon? _____

5. Match each person to a contribution to the development of our understand-

   ing of the solar system.

   ____ (a)   Described a geocentric view of the   (1) Copernicus.
              universe in the *Almagest* about A.D. 150.   (2) Galileo.
   ____ (b)   Determined his three laws of planetary   (3) Kepler.
              motion empirically from observational   (4) Newton.
              data.   (5) Ptolemy.
   ____ (c)   First used a telescope for astronomical   (6) Tycho Brahe.
              work and discovered the phases of
              Venus.
   ____ (d)   Wrote a book describing a helio-
              centric model of planetary motions,
              which was published in 1543, the year
              he died.
   ____ (e)   Formulated the three fundamental laws
              of motion and the universal law of
              gravitation.
   ____ (f)   Observed and recorded planetary motions
              for almost 20 years.

6. What keeps planets in their orbits around the Sun? _____

_____

_____

7. Refer to Figure 8.17. Identify the following points: (a) Sun _____ ;
(b) ellipse _____ ; (c) aphelion _____ ; (d) perihelion _____ ;
(e) force of gravity is greatest _____ ; (f) planet that moves the
slowest _____ .

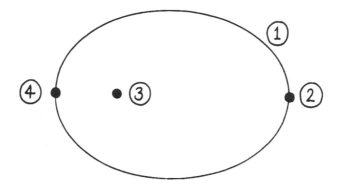

Figure 8.17. Planetary motion.

8. By how much do the Moon's sidereal month and synodic month differ?
Explain. _____

_____

_____

9. What force keeps spacecraft in their trajectories as they travel through the
solar system? _____

10. Classify each of the following as a property of (1) terrestrial planets or (2)
giant planets.

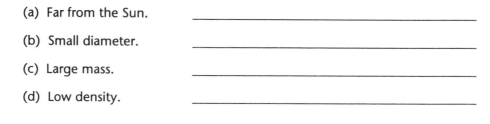

(a) Far from the Sun. _____

(b) Small diameter. _____

(c) Large mass. _____

(d) Low density. _____

(e) Short period of revolution. _____

(f) Short period of rotation. _____

(g) Many moons. _____

11. Match a planet to the appropriate description. *Hint*: Refer to Table 8.2.

____ (a) Closest to Sun.

____ (b) Orbit most inclined to the ecliptic plane.

____ (c) Has the longest sidereal day.

____ (d) Has a year approximately equal to 2 Earth-years.

____ (e) Most massive.

____ (f) Most dense.

(1) Mercury.
(2) Venus.
(3) Earth.
(4) Mars.
(5) Jupiter.
(6) Saturn.
(7) Uranus.
(8) Neptune.
(9) Pluto.

12. What are asteroids? _____

_____

_____

ANSWERS

Compare your answers to the questions on the self-test with the answers given below. If all of your answers are correct, you are ready to go on to the next chapter. If you missed any questions, review the sections indicated in parentheses following the answer. If you missed several questions, you should probably reread the entire chapter carefully.

1. One star, our Sun, orbited by nine planets with their moons, asteroids, comets, and interplanetary gas and dust.   (Section 8.1)

2. Mass. A planet is much less massive and colder than a star. While a star generates its own light, a planet shines by reflecting light from a star. (Section 8.1)

3. All of the planets revolve around the Sun in the same direction. The orbits of all the planets except Pluto lie nearly in the ecliptic plane.   (Section 8.2)

4. (a) Full Moon;   (b) first quarter.   (Section 8.4)

5. (a) 5;   (b) 3;   (c) 2;   (d) 1;   (e) 4;   (f) 6.   (Sections 8.6 through 8.9)

6. A combination of their forward motion and their motion in toward the Sun under the Sun's gravity.   (Section 8.9)

7. (a) 3;   (b) 1;   (c) 2;   (d) 4;   (e) 4;   (f) 2.   (Sections 8.8, 8.9)

8. Two days. While the Moon revolves around Earth, both Earth and the Moon revolve together around the Sun.   (Sections 8.10, 8.11)

9. Gravity.   (Section 8.12)

10. (a) 2;   (b) 1;   (c) 2;   (d) 2;   (e) 1;   (f) 2;   (g) 2.   (Sections 8.13 through 8.15; Table 8.2)

11. (a) 1;   (b) 9;   (c) 2;   (d) 4;   (e) 5;   (f) 3.   (Sections 8.2, 8.13, 8.14; Table 8.2)

12. Irregular, rocky bodies that orbit the Sun, mostly between the orbits of Mars and Jupiter.   (Sections 8.1, 8.16)

9

THE PLANETS

The Earth is the cradle of mankind, but one does not live in the cradle forever.

Konstantin Tsiolkovsky (1857–1935)

Objectives

☆ Compare and contrast the general properties and surface conditions of Mercury, Venus, Earth, and Mars.

☆ Explain what is meant by "morning star" and "evening star."

☆ Compare and contrast the atmospheres of Mercury, Venus, Earth, Mars, Jupiter, Saturn, Uranus, Neptune, and Pluto.

☆ Describe conditions on Mars at the Viking 1 and 2 landing sites.

☆ Give two observations that indicate water might once have flowed on Mars.

☆ Compare and contrast the internal structure of Earth and Jupiter.

☆ Explain the theory of plate tectonics (continental drift) in relation to Earth's geological activity.

☆ Outline two environmental concerns related to Earth's atmosphere.

☆ List and give the current explanation for a famous feature visible in a small telescope for Venus, Mars, Jupiter, and Saturn.

☆ Compare and contrast the general properties of Jupiter, Saturn, Uranus, and Neptune.

☆ Tell what is known about the satellites of Mars, Jupiter, Saturn, Uranus, Neptune, and Pluto.

☆ Describe known properties of Pluto.

9.1 PLANET MERCURY

Mercury, the planet closest to the Sun, is often hidden in the Sun's glare (Figure 9.1). Mercury was appropriately named for the swift Roman messen-

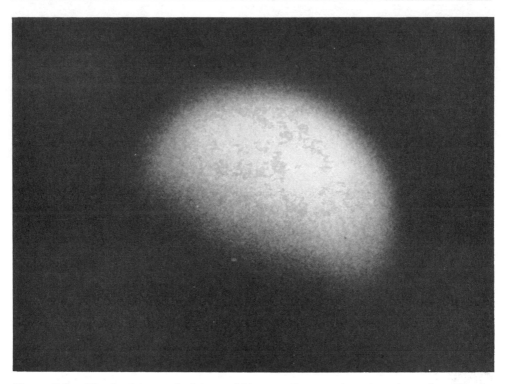

Figure 9.1. The best type of picture of Mercury that can be taken through an Earth-based telescope.

ger god. It wings around the Sun fastest of all the planets, at a mean speed of 172,000 km/hour (107,000 miles per hour).

Our first close-up views of Mercury came from the U.S. robot Mariner 10, which photographed half of the planet on three flybys in 1974–1975 (Figure 9.2).

Mercury looks like our Moon. The surface is ancient and heavily cratered. The craters suggest that meteorites bombarded the inner planets in the final stages of their formation. The largest crater, Caloris Basin, is 1300 km (800 miles) across. Large, smooth areas resembling the Moon's maria suggest that extensive lava flooding occurred in the past.

Huge **scarps**, or cliffs, up to 2 km (1.2 miles) high and 1500 km (930 miles) long crisscross the planet. They apparently formed when Mercury's interior cooled and shrank, compressing the crust.

Mercury's axis of rotation is vertical (not tilted as Earth's is), so the Sun is always exactly overhead on the equator there. The planet has no seasons, and some sunlight always shines on its poles.

Temperatures vary from extremely hot in direct sunlight, 700 K (430° C or 800° F), to bitter cold, 90 K (–180° C or –300° F), on the dark side.

The very high temperatures release volatile substances from the surface, which creates an extremely thin atmosphere. Helium, sodium, hydrogen, and possibly oxygen have been detected. The surface air pressure is barely 2 trillionths of Earth's at sea level ($<2 \times 10^{-9}$ millibars). Most of the gases, includ-

Figure 9.2. Mercury—a mosaic of over 200 pictures taken by Mariner 10. The largest craters are about 170 km (100 miles) in diameter. Bright rays from fresh impact craters, smooth dark plains, and large scarps are also visible.

ing possible water vapor, escape from the planet, but some may be trapped by the cold regions at the poles and precipitate as a kind of snow.

Mercury's unphotographed hemisphere was first imaged by Earth-based radar in 1991. Variations in the surface reflectivity of radar signals reveal the topography. Results suggest that there are ice deposits at Mercury's poles. The temperature at the polar regions seems to be cold enough to retain ice—125 K (–148° C or –235° F). Mercury has a very weak magnetic field that affects the moving charged particles in the solar wind. It could also affect the observed variable sodium abundance in the thin atmosphere.

If Mercury has a crust of light silicate rock (average density about three times that of water), how can you explain Mercury's average density of almost 5.5 times that of water? _____

_____

Answer: Mercury probably has a very dense core. (Scientists figure that Mercury has a dense iron core about the size of our Moon, surrounded by a rocky crust.)

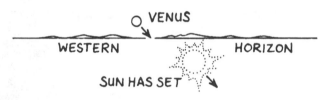

Figure 9.3. Venus as an evening star in the western sky right after sunset.

9.2 VENUS: OBSERVING

Brilliant **Venus** was named for the Roman goddess of love and beauty. At night Venus outshines all the stars. The planet is so conspicuous that it is often mistakenly reported as an unidentified flying object (UFO) (Figures 9.3 and 9.4).

Venus, like Mercury, circles the Sun inside Earth's orbit. As a result, both planets appear close to the Sun in our sky. They shine in the western sky just after sunset near their eastern elongation. Then they appear to follow the Sun across the sky. They are frequently called **evening stars** at that time.

They are **morning stars** in the eastern sky just before sunrise near their western elongation. Then they lead the Sun across the sky.

Both Venus and Mercury go through a cycle of phases (Figure 8.7) that you can observe in a small telescope. Venus rotates from east to west, or **retrograde**.

Normally Venus and Mercury pass above or below the Sun at conjunctions. About 13 times in a century Mercury, and much less frequently Venus, **transit**, or pass directly in front of, the Sun, at conjunction. Observers see a tiny dot moving across the bright face of the Sun. Mercury will transit on November 15, 1999, and Venus on June 8, 2004, and June 5, 2012.

Venus is at inferior conjunction at intervals of 584 days. Then it comes closer to Earth, about 40 million km (25 million miles) away, then any other planet.

Refer to Figure 9.5. Determine the location of Venus when it is (a) an evening star _____ ; (b) a morning star _____ ; (c) at conjunction _____

Answer: (a) 1; (b) 2; (c) 3.

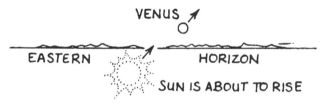
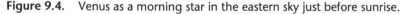

Figure 9.4. Venus as a morning star in the eastern sky just before sunrise.

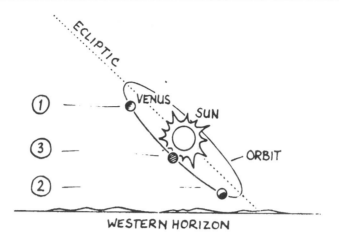

Figure 9.5. Venus in orbit.

9.3 VENUS: THE PLANET

Venus shines brilliantly because it is shrouded in thick clouds that reflect a lot of sunlight. These perpetual clouds hide the surface from our view.

More than 20 U.S. and Russian robot spacecraft have successfully encountered Venus and transmitted data back to Earth for analysis and image processing. Spacecraft firsts at Venus include Mariner 2/flyby 1962, Venera 3/impact 1965, Venera 7/soft landing 1970, and Venera 9/orbit 1975 (Figure 9.6).

The atmosphere is about 97 percent carbon dioxide and 1 to 3 percent nitrogen with traces of water vapor, helium, neon, argon, sulfur compounds, and oxygen. It circulates in large global motions. The temperature of the cloud tops is about 250 K (–9° F). Cloud layers about 19 km (12 miles) thick (altogether) are about 50–70 km (40 miles) above the surface. Apparently they are colored yellow by corrosive sulfuric acid.

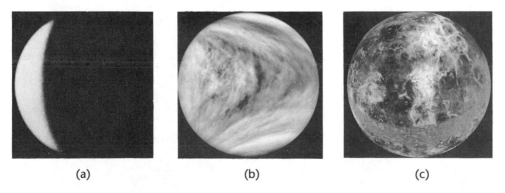

Figure 9.6. Venus as viewed by (a) Earth-based telescope. (b) U.S. Pioneer Venus Orbiter. (c) Computer-generated map from Magellan radar imagery and altimetry data.

Figure 9.7   The first look at the surface of Venus, from the short-lived Venera spacecraft.

Venera landers first took surface pictures in 1975. Rocks and soil look orange beneath the thick clouds. In direct sunlight on Earth, they would look gray. The landers found a very inhospitable world. They all expired within two hours because of hellish conditions (Figure 9.7).

Surface temperatures reach 755 K (482° C or 900° F) because carbon dioxide and water vapor in the clouds let in visible light from the Sun but do not let out infrared heat that is given off by the hot rocky surface. This **green-**

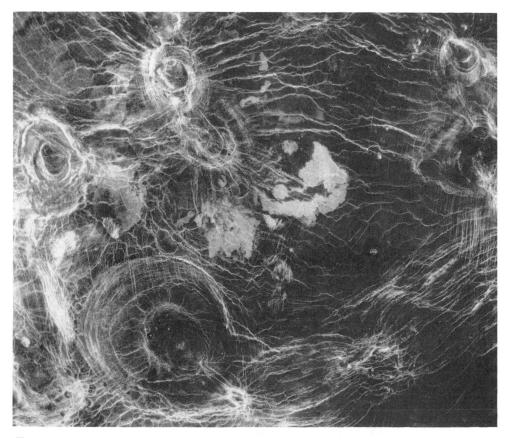

Figure 9.8.   Arachnoids, found only on Venus. These are approximately 50 km to 230 km in diameter, on radar-dark plains on Venus.

house effect makes Venus ever hotter. Atmospheric pressure is a crushing 90 times more than Earth's (over 90 atmospheres, or 1330 pounds per square inch). There are many lightning bolts and thunderclaps.

To map the surface, scientists bounce radar signals from Earth or robot spacecraft off Venus and analyze the echoes. Our best radar imaging data came from the U.S. Venus orbiter Magellan (1989–1994) which mapped 99 percent of the planet (Figures 9.8 and 9.9) and also studied the interior and atmosphere.

Radar images show that all the terrain is dry and rocky. About 80 percent is relatively flat plains with fractures, impact craters, and volcanoes that are within 1 km of the planet's mean surface. The difference between the lowest and highest elevations is 15 km (9 miles).

Surface features on solar system bodies are customarily given two names by the International Astronomical Union. One is descriptive and is used for geographic features on all bodies. The other name is for identification. On Venus names of surface features honor love goddesses or famous women who have been dead for at least three years. The sole exception is Maxwell Montes, the largest mountain, which honors the Scottish physicist James Clerk Maxwell (1831–1879), whose theories of electromagnetism made radar possible.

Impact craters on Venus look different from those on other worlds. Small

Figure 9.9. Western Eistla Regio scene developed from Magellan radar imagery and altimetry data. A rift valley in the foreground extends to the base of Gula Mons, a 3-km-high volcano at the right. Sif Mons, a 300-km-wide and 2-km-high volcano, is at the left.

chunks of rock from space burn up in the thick atmosphere. Large impactors crash onto Venus at about 20 km (12 miles) per second, releasing a tremendous amount of energy which vaporizes the crashing object and the surrounding ground. **Ejecta**, surface material, is blasted out. It stays molten because of the very high surface temperature and flows away from the crater in patterns that look like flower petals.

Highlands that appear like continents tower above the dry plains. Terra Aphrodite, the largest, is half the size of Africa. Smaller Ishtar Terra is the size of the continental United States. Here Maxwell Montes, a mountain massif, soars nearly 11 km (7 miles) above the mean radius. There are many fault zones.

Numerous strange volcanic features dot the plains. **Shield volcanoes**, small domes 2–3 km (1–2 miles) across and a few hundred meters high, are caused by a thin, runny lava that builds a broad mountain. **Pancake domes**, typically 25 km (15 miles) wide and less than a mile high, are associated with a thick, sticky magma with more than the normal silica content. **Coronae**, circular rings of low-relief folds, were most likely formed when lava flows created a dome, which then sank and collapsed. Some volcanoes may still be active.

Venis is close to Earth in size, mass, density, and distance from the Sun. However, you could not live there comfortably. Give three reasons why not.

(1) _____ ;

(2) _____ ;

(3) _____

Answer: Venus (1) is much too hot: 482° C (900° F); (2) has a poisonous carbon dioxide atmosphere; and (3) has a crushing atmospheric pressure, over 90 times Earth's.

9.4 PLANET EARTH

Our own planet **Earth** shines like a rare blue and white jewel in space (Figure 9.10). Third from the Sun, it is the most important planet of all to us.

The total surface area of our planet is almost 5.10×10^8 km² (199 million square miles. More than 70 percent of our planet is covered by water, which is unique in the solar system.

The highest mountain on Earth is Mt. Everest in Asia, almost 9 km (29,000 feet) above sea level. The deepest measured underwater spot is the Marianas Trench, more than 11 km (36,000 feet) below the Pacific Ocean's surface.

Earth's mass is about 6×10^{24} kg. This mass provides the surface gravity we are used to.

Figure 9.10. Earth viewed from the Moon by Apollo astronauts.

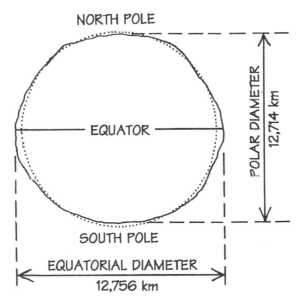

Figure 9.11. Earth's daily rotation around its axis has produced an equatorial bulge and polar flattening (exaggerated).

Refer to Figure 9.11. How much longer in kilometers is the distance across Earth's equator than the distance from its North to South Pole? _____

Answer: About 43 km.

Solution:

Equatorial diameter – Polar diameter = 12,756.34 km – 12,713.80 km = 42.54 km.

9.5 EARTH'S STRUCTURE

Astronomers figure that Earth was born about 4.6 billion years ago. It formed together with the other planets out of the same contracting cloud of gas and dust that formed the Sun (Section 4.3).

Geologists cannot go inside Earth to examine it directly. Instead, they determine its structure and composition from the way seismic waves, from earthquakes and explosions, are transmitted through the Earth and along its surface. They picture Earth today in three layers: The **crust** is the thin, outermost, solid layer. It is an average of 35 km (22 miles) thick, being thicker where there are continents and thinner under the oceans. The crust is composed mainly of lightweight rocks such as granite and basalt. The **mantle** is the layer

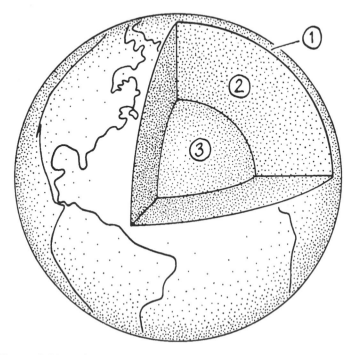

Figure 9.12.   The structure of Earth showing three principal layers.

below the crust. It extends in about 2880 km (1,800) miles. Laboratory analysis of samples from volcanoes indicates that the thick mantle consists mostly of dense silicate rock that behaves somewhat like taffy—yielding under steady pressure but fracturing under impact. The central layer, which is 3470 km (2170 miles) thick, is called the **core**. Here an outer, molten, metallic layer about 2100 km (1300 miles) thick probably surrounds a solid center. The core is probably made of dense iron and nickel at a temperature of about 6400 K.

Refer to Figure 9.12. Identify the three principal layers of Earth and state the approximate thickness of each. (1) _____ ;

(2) _____ ; (3) _____

Answer: (1) Crust—an average of 35 km (22 miles); (2) mantle—about 2880 km (1800 miles); (3) core—about 3470 km (2170 miles).

9.6 EARTH'S GEOLOGICAL ACTIVITY

The surface of our restless Earth is constantly changing because of erosion and geological activity. The oldest rocks discovered so far amid the remote lakes and tundras of northwest Canada are about 3.96 billion years old.

Substantial evidence indicates that about 200 million years ago all of the world's continents were joined in one huge supercontinent called Pangaea, which later broke up.

According to the theory of **plate tectonics**, also called the **continental drift theory**, continents and ocean floor are embedded in plates, or rock slabs, several thousand miles across. The plates move slowly on the slightly yielding mantle beneath. Earth's crust is reshaped at plate boundaries. Where the plates move apart, the continents separate slowly, at about 2.5 cm (1 inch) per year. That adds up to over 5000 km (3000 miles) in 200 million years.

Movement of the plates is also responsible for mountain building, earthquakes, and volcanic activity. These events occur at boundaries between the moving plates where they press on each other forcibly.

A popular theory says that magma convection currents power the continental drift. Magma currents flow up through the mantle. Upon meeting cool, rigid rocks, they flow horizontally. Friction drags the continent-bearing plates along. Finally the cooled magma sinks. Along midocean ridges, magma pours through the crust continually creating new rocks.

Remarkable evidence confirms that the continents drift. Laser-ranging devices bounce light pulses off reflectors on the Moon and measure drift. The coastlines of South America and West Africa seem to fit together, and similar plant and animal fossils are found along both, although they are now separated by almost 5000 km (3000 miles) of Atlantic Ocean. (Figure 9.13).

The ages of rocks from the bottom of the Atlantic Ocean have been measured. The oldest, found near the continent coastlines, are 150 million years old.

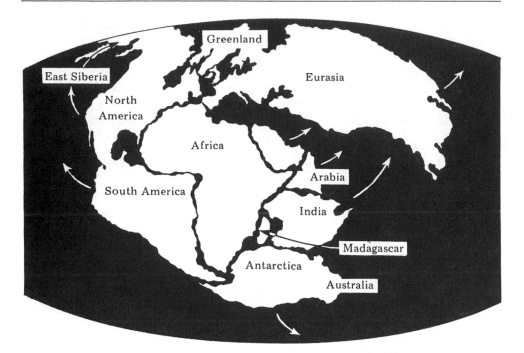

Figure 9.13. A map of Earth as it probably appeared about 200 million years ago.

If 4-billion-year-old rocks had been found on the bottom of the Atlantic Ocean, how would the continental drift theory have been affected? Explain.

_____

_____

_____

_____

Answer: Serious doubts would have been raised about the correctness of the theory, which says that the Atlantic Ocean, almost 5000 km (3000 miles) wide between the continental coastlines, formed in the last 200 million years. It did not exist 4 billion years ago.

9.7 EARTH'S MAGNETISM

Our planet has a **magnetic field**, or region of magnetic forces, that affects compass needles.

The magnetic north pole is located at about 76°N latitude, 101°W longitude in northeast Canada. It is about 1300 km (800 miles) from the geographic North Pole and slowly changes position as time goes by. The magnetic south pole is located at about 66°S, 140°E. Earth's magnetic field is believed to be generated by its liquid iron-nickel core, which acts like a giant dynamo as the planet spins. The complex motion of this core probably causes the long-term migration of the magnetic poles.

Earth's **magnetosphere**, the region around the planet where its magnetic field is influential, extends out into space about 4 Earth radii on the sunward side. The **magnetotail**, the part of the magnetosphere on the side away from the Sun, extends like a tail from 10 to perhaps 1000 Earth radii.

Many energetic, charged particles from the solar wind that could be deadly are trapped by Earth's magnetic field. They keep moving around rapidly inside two doughnut-shaped regions called the **Van Allen belts** in the magnetosphere.

What is the magnetosphere? _____

_____

Answer: The region surrounding Earth where the magnetic field is influential.

9.8 EARTH'S ATMOSPHERE

Earth is surrounded by an **atmosphere** that extends several hundred miles out into space.

Earth's first atmosphere over 4 billion years ago was probably very different from air today. Noxious compounds of hydrogen, carbon, oxygen, and nitrogen—such as carbon dioxide, ammonia, and methane, plus water vapor—may have **outgassed**, or been released from the interior of the hot, young planet.

The carbon dioxide, which is highly soluble in water, could have been removed by combining with substances such as calcium in the ocean to form limestone. The free oxygen we need for respiration probably was generated by green plants. In photosynthesis, green plants absorb carbon dioxide from the air, utilize it for growth, and release oxygen.

Air today contains about 78 percent nitrogen, 21 percent oxygen, and 1 percent argon, carbon dioxide, and other gases. It also has variable amounts of water vapor, dust, carbon monoxide, chemical products of industry, and microorganisms.

Over half of this air is packed within the first 6 km (4 miles) above Earth's surface. The air thins out fast with increasing altitude. At about 12–50 km (7–30 miles) above sea level, the Sun's ultraviolet light acts on air to produce **ozone**, a molecule made up of three atoms of oxygen. A global ozone layer shields people and plants from harmful ultraviolet radiation from the Sun. Beyond 160 km (100 miles), satellites can orbit without being dragged down.

Researchers are using sophisticated computer simulations and instruments on the ground and aboard airplanes and spacecraft to study potentially dangerous changes in the atmosphere and climate caused by human activity. Elevated levels of ozone-destroying chlorofluorocarbons released by refrigerators, air conditioners, and some aerosol sprays may cause the observed gaping holes in the ozone layer over the polar regions and thinning over the midlatitudes. Increasing concentrations of carbon dioxide and impurities released by burning coal and oil and the clearing away of rain forests might cause a global warming of our planet similar to the greenhouse effect on Venus.

The total mass of the entire atmosphere is about 5000 trillion tons. Gravity keeps the atmosphere tied to Earth, although atoms occasionally escape at the top. At sea level all that air presses down with a force of 1.03 kg/cm^2, (14.7 pounds per square inch), called 1 atmosphere of pressure. The millibar is another common unit of atmospheric pressure. At sea level, the air pressure on Earth is about 1013 millibars.

What is the (a) composition and (b) pressure at sea level of the air that supports our lives on Earth? (a) _____

_____

(b) _____

_____

Answer: (a) About 78 percent nitrogen, 21 percent oxygen, and 1 percent carbon dioxide and other gases; variable amounts of water vapor and impurities. (b) About 1.03 kg/cm^2 (14.7 pounds per square inch), also called 1 atmosphere and 1013 millibars.

9.9 MARS: OBSERVING

Red **Mars** reminded the Romans of blood and fire, so they named the planet after their god of war. It has two small moons, appropriately called Phobos ("fear") and Deimos ("terror"), which can be seen only in powerful telescopes.

Superior planets like Mars look brightest when they are on the opposite side of Earth from the Sun, a position called **opposition**. Then a fully lighted disk faces us. Mars is at opposition at intervals of 780 days on the average.

Superior planets are hardest to observe when they are on the opposite side of the Sun from Earth, a position called conjunction Figure 9.14).

Mars comes closer to Earth at some oppositions than it does at others because of the eccentricity of its orbit (Table 8.1). Close oppositions are called **favorable**, because the disk of Mars looks larger, and observing is better. The most favorable oppositions occur when Mars is near perihelion (Figure 9.15). Then Mars is only about 56 million km (35 million miles) away from Earth. This happens toward late summer at intervals of 15 to 17 years.

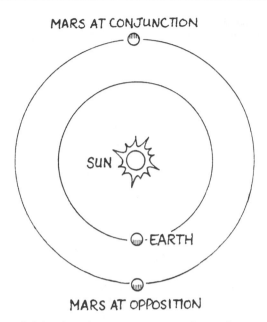

Figure 9.14.   Two important aspects of Mars from Earth.

When Mars is near **favorable opposition** you can, with a telescope, see features that have long excited the imagination. In each hemisphere, white polar caps, which shrink during the summer, are visible. Dust storms and dark areas once mistakenly thought to be water or vegetation may also be seen. The dark areas are probably surface exposed after dust storms (Figure 9.16).

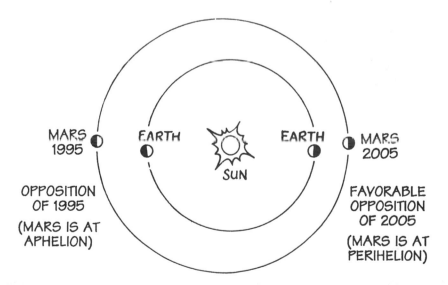

Figure 9.15.   Favorable and unfavorable oppositions of Mars in the current cycle.

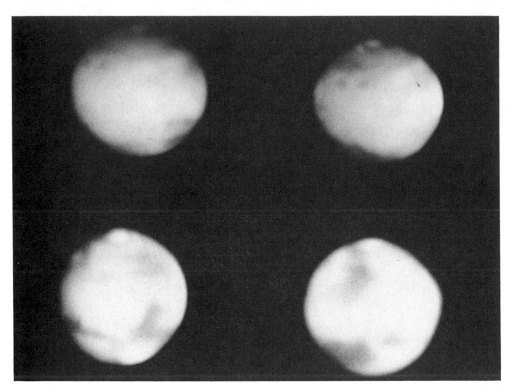

Figure 9.16. Four photographs of the planet Mars during its 1967 opposition. The photographs were taken with the Catalina Observatory 61-inch telescope in April and May 1967.

"Canali," straight, dark channels, were first reported in 1877 by Italian astronomer G. V. Schiaparelli. This word was mistranslated into English as "canals". U.S. astronomer Percival Lowell (1855–1916) caused great excitement at the beginning of the twentieth century when he mistakenly said that intelligent Martians had built the canals. Eight successful U.S. and Russian probes, starting with U.S. Mariner 4 in 1965, found no canals on Mars. Some of these markings may actually be mountain ranges.

Why is Mars best observed at favorable oppositions? _____

_____

Answer: It is closest to Earth then. (Both Mars and Earth travel in elliptical orbits around the Sun, so the distance between them varies considerably.)

9.10 MARS: THE SURFACE

We took our first good look at the surface of another planet through the "eyes" of the U.S. robot Viking Lander 1, which set down on Mars on July 20, 1976,

Figure 9.17. Historic first 300° panoramas of the surface of Mars from Vikings 1 and 2.

seven years to the day after Apollo astronauts first walked on the Moon (Figure 9.17).

Scattered rocks, sand dunes, and distant hills came into view when Viking Lander 1 settled into the powdery dirt in a plain called Chryse Planitia (Plains of Gold). The site is located at 22.46°N latitude, 48.01°W longitude on Mars. Air temperature ranged from –86° C (–122° F) shortly after dawn to a high of –30° C (–22° F) at midafternoon on that Martian summer day. Air pressure was only about 7 or 8 millibars.

Two months later Viking Lander 2 set down in a plain called Utopia Planitia (Utopia Plains) at 47.89°N latitude, 225.86°W longitude, about 7500 km (4600 miles) northwest of its predecessor. The site is marked by pitted rocks that look like those produced by gaseous volcanoes or meteorite impacts on Earth.

The Viking landers sent back over 4500 pictures of the Martian surface, 3 million weather reports, and data from chemical and biological tests. Viking Lander 1 sent weekly pictures and weather reports to Earth until November 13, 1982. In 1984 this historic robot became the first museum exhibit located on another world when ownership was transferred to the U.S. National Air and Space Museum.

The rusty red soil at the landing sites looks like iron-rich clay. Rocks are covered with fine-grained reddish material. Chemical weathering of the rocks and erosion seem to have occurred. Lander tests found that the fine soil is about 45 percent silicon oxide and 19 percent hydrated iron oxide (rust). The sky is colored pink in daytime by red dust that hangs in the atmosphere like smog. Sunsets are pale blue. In the winter temperatures drop below –123° C (–190° F), and a fine layer of frost appears.

No signs of large organisms were apparent at either site. There was no flow-ing water. Both landers had been sterilized before launch so they would not contaminate either Mars or their own biology experiments with organisms brought from Earth. Results of tests for microorganisms in the soil remain puzzling.

Briefly describe the surface of Mars at the Viking lander sites. _____

_____

_____

Answer: It looks like a red, dry, rock-strewn desert. The sky is pink. The temperature is cold.

9.11 MARS: THE PLANET

Viking 1 and 2 consisted of orbiters as well as landers. The orbiters took over 51,500 pictures of Mars while orbiting the planet. They mapped 97 percent of its surface with a resolution of 300 m (1000 feet) and 25 percent with a resolution of 25 m (82 feet) or better.

Viking Orbiters 1 and 2 showed us a harsh, rugged, dry planet. The northern hemisphere is basically plains and the south is mostly cratered. Polar ice caps and great volcanoes are prominent features.

Mars has huge volcanoes, some of which may still be active. Olympus Mons (Mount Olympus) is the largest volcano in the solar system. It towers almost 27 km (17 miles) above the mean surface and contains more lava than the U.S. Hawaiian Islands.

Gravity data, derived from changes in the orbits of the Vikings, and extremely tall, massive volcanoes indicate that the planet's crust is about 50 km (30 miles) thick and does not drift as Earth's continents do. Mars probably has a mantle that is cooler and thicker than Earth's, and a molten iron core.

The planet has deep canyons. The largest, Valles Marineris (Mariner Valley), is a complex network of rocky valleys extending 5000 km (3000 miles) around the equator with an average depth of 6 km (4 miles). Diverse areas evidence great landslides, faulting, and flow channels carved deeper by erosion.

Craters suggest that meteorites bombarded the planet. The lowest point on Mars, 6 km (4 miles) below the mean surface, is the bottom of the circular Hellas Planitia basin. Hellas was apparently created by a meteor or asteroid impact about 4 billion years ago. Younger craters, such as the 18-km (11-mile) wide Crater Yuty, look as if water and shattered rock poured out and flowed for great distances after a big impact.

Mars does not have liquid surface water today. However, there is indirect evidence of ancient catastrophic flooding. Its deep, winding channels resemble river beds with tributaries (Figure 9.18). They look as if they were carved long ago by great rivers. That water may be locked in the ice caps and permafrost under the surface. Climate changes may have turned a running water environment into the cold, dry world Mars is today.

Figure 9.18. Deep, winding channels and craters on Mars from Viking Orbiter.

Water is evident in solid and vapor form. The permanent ice cap at the north pole is made of frozen water. It is covered in winter by layers of carbon dioxide that freezes out of the atmosphere. The ice cap at the south pole is frozen carbon dioxide. Winter frost is apparently frozen water and dust. There are also occasional fog and filmy clouds.

The atmosphere is too thin to block the deadly ultraviolet rays from the Sun that beat down on the planet. It is made up of about 95 percent carbon dioxide, with 2 to 3 percent nitrogen, 1 to 2 percent argon, and 0.1 to 0.4 percent oxygen, with traces of water vapor and other gases.

Wild dust storms swirl out of the southern hemisphere in the summer. Often they rage over the whole planet. Winds up to 120 km (75 miles) per hour blow light-colored dust about, sculpting and exposing dark rock. Thin layers of ice and dust hundreds of kilometers long have been laid down at the north and south poles by global dust storms in alternating seasons.

A piloted flight to Mars could realistically take place early in the twenty-first century. First, the U.S. expects robots to send us mapping and surface data for one Martian year. Further exploration by robots, including a sample return mission, are proposed for this decade.

Scientists think water is critical for life. Perhaps life formed on Mars in the distant past when that planet was warmer and wetter. Possibly microbes still survive.

List two pieces of evidence that indicate water once flowed on Mars.

(1) _____

_____

(2) _____

_____

Answer: (1) The deep, winding surface channels look as if they were carved by great rivers. (2) The permanent polar ice cap at the north pole is made of frozen water that may once have flowed on Mars.

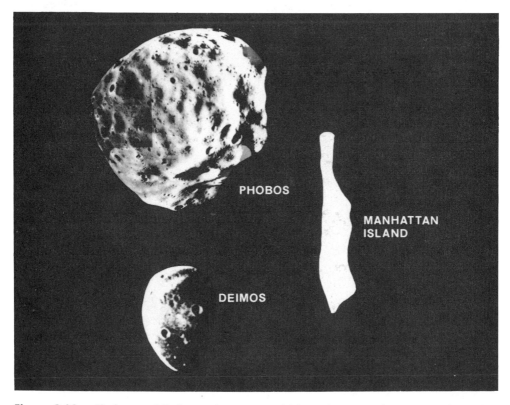

Figure 9.19. Phobos and Deimos, the moons of Mars, shown at the same scale as the island of Manhattan.

9.12 MOONS OF MARS

Phobos and Deimos are small, irregular rock chunks only about 21 km (13 miles) and 12 km (7 miles) long, respectively. (Figure 9.19). Phobos orbits Mars every 7.7 hours, while Deimos completes a circuit in 1.3 days.

Both moons look fairly old, with many impact craters of varying ages. Phobos has striations and chains of small craters. Stickney, its largest crater, measures practically 10 km (6 miles) across.

Two moons of Mars were mentioned by English writer Jonathan Swift in 1727, long before they were actually discovered in 1877 by U.S. astronomer Asaph Hall (1829–1907).

Briefly describe the moons of Mars. _____

Answer: Small, irregular, cratered rock chunks.

9.13 JUPITER: OBSERVING

Giant **Jupiter** was named for the mythological Roman king of the gods and ruler of the universe. The biggest planet of all, Jupiter at night outshines the stars and all the planets except Venus. Jupiter's colorful, parallel, dark and light cloud bands, Great Red Spot, and four biggest moons can be seen through a small telescope. Discovered by Galileo and collectively called the **Galilean moons**, Io, Europa, Ganymede, and Callisto change patterns nightly as they circle the planet. Astronomical publications and computer software (see Useful Resources) list current moon positions, occultations, and transits.

Our best look at the Jovian system came from the U.S. robots Voyagers 1 and 2 encounters in 1979. Voyager 1 flew within 206,700 km (128,400 miles) and Voyager 2 within 570,000 km (350,000 miles) of Jupiter's cloud tops. The two spacecraft sent back more than 33,000 pictures.

The U.S. robot Galileo is en route to explore the Jovian system in 1995. Galileo is to split into two parts near Jupiter. An atmospheric probe is to plunge through Jupiter's clouds and transmit data for an hour before heat and pressure kill it. An orbiter is to collect and transmit data on Jupiter and its moons for two years.

Through a small telescope Jupiter's four brightest moons look like stars. What observations show they are really satellites of the planet? _____

Answer: The moons are observed to change positions nightly as they circle the planet.

9.14 JUPITER: THE PLANET

Jupiter is more massive than all the other planets and their moons combined. It just missed being a star. If Jupiter were about 80 times more massive, nuclear fusion reactions could have started. (Figure 9.20).

The planet seems to be a huge, rapidly spinning, liquid ball topped by a thick atmosphere, made mostly of hydrogen and helium. Apparently it has a relatively small solid core. A faint, thin ring system of particles (from microscopic to a few meters in size) encircles Jupiter. The outermost part, a gossamer ring just beyond a brighter one, extends to some 210,000 km (130,000 miles) from the planet's center.

Colorful changing cloud features and convoluted weather patterns circulate in the dynamic, observable atmosphere. Superbolts of lightning flash. Complex patterns show in and between the moving dark-colored **belts** and **lighter zones.** Hydrogen, helium, the detected traces of methane, and water vapor are all colorless. Sulfur or phosphorous compounds and ammonia at various depths must give the atmosphere its bright red, orange, yellow, and brown colors and white clouds. The famous **Great Red Spot** is a colossal atmospheric storm. It has been observed for over 300 years at varying sizes, brightness, and color. The Great Red Spot rotates counterclockwise and also moves around the planet. Cooler than surrounding clouds, it towers up to 24

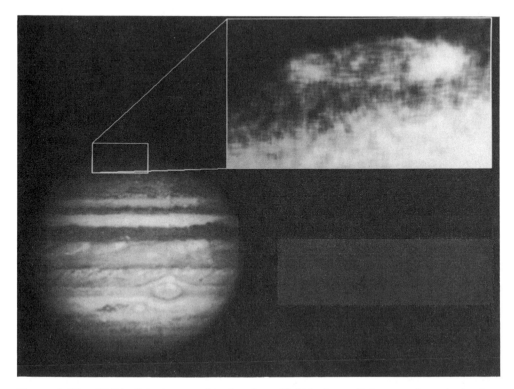

Figure 9.20. Visible-light image of Jupiter from Hubble Space Telescope. Inset is the first direct ultraviolet image of a Jupiter aurora.

km (15 miles) above them. Smaller storms and eddies appear throughout the banded clouds.

Temperatures hit 160 K (–170° F) at the cloud tops. The atmosphere extends down about 1000 km (600 miles). Most likely the density of hydrogen increases steadily from the top inward until it changes to liquid hydrogen. Near the planet's center, the pressure and temperature could be high enough to compress hydrogen to an extraordinarily dense form called liquid metallic hydrogen.

At the core temperatures may be 30,000 K (53,000° F), which would explain the observation that Jupiter radiates about twice as much heat as it receives from the Sun. The planet has a powerful magnetic field that traps ions and electrons in a complex system of large, intense radiation belts. Plasma (a collection of ions and electrons) oscillations account for some of Jupiter's observed radio emission. The magnetic field is essentially dipolar but opposite Earth's in direction. Electrical currents in the liquid hydrogen layer could be its source. At the cloud tops, Jupiter's magnetic field is 1.5 to 7 times more powerful than Earth's. Jupiter's enormous magnetosphere varies in size, possibly due to changes in the solar wind pressure. It may stretch sunward 7 million km (4 million miles) and outward nearly 650 million km (400 million miles) to Saturn's orbit.

Jupiter's atmosphere is especially interesting, because it may be similar to Earth's primitive one.

What is Jupiter's atmosphere made of? _____

Answer: Mostly hydrogen and helium, with traces of methane, ammonia, water vapor, and other gases.

9.15 JUPITER'S MOONS

There are 16 known moons orbiting Jupiter (see Table 8.3). Most are small. Voyagers 1 and 2 focused on the four biggest plus innermost Amalthea and discovered three small moons (Figure 9.21).

Small Amalthea resembles a dark red football. Its elongated surface shows signs of meteorite impacts.

Colorful Io has active volcanoes which spew sulfur-rich materials that color the surface bright orange, red, brown, black, and white. Io's bright white spots are sulphur dioxide frost, and its tenuous atmosphere is primarily sulphur dioxide gas. The volcanoes maybe due to the heating that occurs as Europa and Ganymede tug gravitationally on Io and Jupiter alternately pulls Io back to its regular orbit. This pumping creates tidal bulges on Io's surface that are up to a hundred times greater than the typical 1m (3.3 feet) tidal bulges on Earth.

A gigantic cloud of charged particles, mostly ions of sulphur and oxygen, wobbles around Jupiter at Io's distance. The particles are likely stripped off Io by magnetic forces as Jupiter's magnetosphere rotates with the planet. Cloud particles may also travel along Jupiter's magnetic field lines into its north and south polar atmospheres, causing brilliant Jovian auroras.

There is evidence of water ice on the surfaces of Europa, Ganymede, and

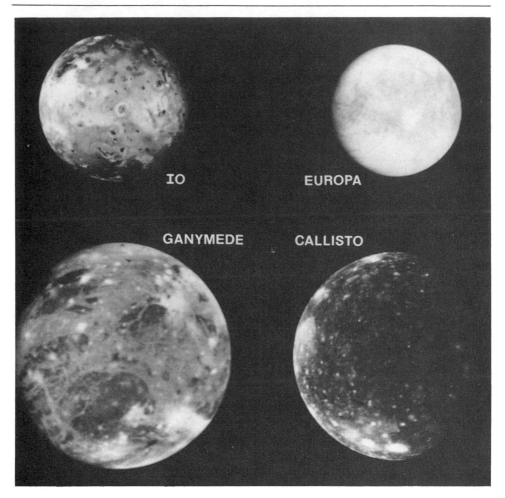

Figure 9.21. Composite photograph of Jupiter's four Galilean moons from Voyager.

Callisto. Europa, about the same size and density as our Moon, is the brightest Galilean moon. Its smooth, icy crust is crisscrossed by puzzling long lines.

Ganymede and Callisto may be composed of up to 50 percent water, mixed with rocky material. Ganymede is the largest known moon in the solar system, 5260 km (3261 miles) in diameter. It has dark, probably ancient, areas with many craters and lighter, younger terrain that is grooved, suggesting global tectonic activity. Callisto's surface looks oldest, with numerous impact craters. The largest craters may have been erased by the flow of icy crust. Features that look like the remains of very large basins may record collisions with large chunks of rock and metal.

(a) What is the largest moon in the solar system? _____ ;

(b) What is its diameter? _____

Answer: (a) Ganymede; (b) 5260 km (3261 miles).

Figure 9.22. The various aspects of Saturn from Earth. Last occurrence of maximum inclination of the north side of the rings toward the Sun, in 1987; south side of the rings toward the Sun, in 1973. The rings were last edge-on in 1980 and will be again in 1995.

9.16 PLANET SATURN

Saturn, the most distant bright planet, was named for the Roman god of agriculture. Dazzling rings surround Saturn (Figure 9.22). They were named in order of their discovery. From the planet outward, they are known as D, C, B, A, F, G, and E.

We see the rings at various angles, from edge-on to 29°, as both Earth and Saturn orbit the Sun. Since Saturn takes 29.5 years to orbit the Sun, we see that planet in essentially the same orientation toward Earth in the same area of our sky for months. Although the brightest rings are 65,000 km (40,000 miles) wide, they are only a few kilometers (miles) thick. Stars can be seen through them (Figure 9.22).

Voyager 1 in 1981 flew within 64,200 km (40,000 miles) and Voyager 2 in 1982 within 41,000 km (26,000 miles) of Saturn's cloud tops. The Voyagers sent back 33,000 images of the Saturnian system.

Saturn's rings consist of dust- to boulder-size icy particles that resemble icy snowballs or ice-frosted rocks orbiting Saturn. They shine by reflecting sunlight. Probably the larger particles are the remnants of moons that were shattered by impacts and the rest resulted from collisions. Or the rings may be material that never collected into a single moon.

Figure 9.23. Rings of Saturn from Voyager, computer enhanced to bring out fine details. The brighter rings are about 5 times wider than Earth (shown to the same scale), but their thickness is barely 100 m.

Hundreds of tiny ringlets comprise the A, B, and C rings, which are seen through small telescopes. Long, spokelike features in the B ring may be shiny, fine particles raised by electrostatic forces.

The F ring was discovered by the U.S. robot Pioneer 11 in 1979. It has separate ringlets, which are partly intertwined and kinked by the gravitational forces of two small moons that shepherd the ring material. Voyager 1 confirmed the existence of the D and E rings, and discovered the G ring. (Figure 9.23).

Like Jupiter, Saturn is a huge, multilayered gas ball with perhaps a relatively small iron-silicate core. It has a dynamic atmosphere that is flattened at its poles by rapid rotation. Colors and features such as belts and zones as well as long-lived ovals are much less distinct because of a hazy layer above the visible clouds. Saturn's atmosphere has the same constituents as Jupiter's, but in different mixtures. It contains less than half the amount of helium. Saturn also radiates more energy than it absorbs from the Sun. Perhaps the precipitation of helium out of the mainly hydrogen atmosphere could be supplying Saturn's internal heat.

At about 29.5-year intervals, when Saturn's northern hemisphere receives the most heat from the Sun, a large white spot suddenly appears. The spot, thousands of kilometers wide, is a giant storm of gas rising up from deep in the atmosphere. The Great White Oval of 1990 appeared on September 24, spread far around the planet's equatorial zone in October, and faded from view in November.

The highest-speed winds, over 1600 km (1000 miles) per hour, occur at the equator and are much stronger than Jupiter's. Temperatures near the cloud tops range from 86 K (–305° F) near the center of the equatorial zone to 92 K (–294° F). There are auroral emissions and lightning.

With a mass equal to 95 Earths in a volume 844 times Earth's, Saturn has the lowest average density of all the planets. It could float in water if a big enough sea existed anywhere.

Saturn's magnetosphere is about one third as big as Jupiter's. It, too, varies in size when the solar wind intensity changes. It may extend sunward nearly 2 million km (1 million miles). The magnetic field drags along charged particles that circle Saturn as the planet rotates.

U.S. and European scientists propose to send a spacecraft named Cassini to explore Saturn in the twenty-first century. Near Saturn, Cassini would split into two parts. A probe would perform tests and snap pictures while descending through the atmosphere of the moon Titan. If the probe landed safely, it would send a brief surface report. An orbiter would radio back data on Saturn and its satellites for four years.

(a) What are Saturn's rings made of? _____

_____

(b) Can you explain why they look solid in a small telescope? _____

_____

Answer: (a) Dust- to boulder-size icy particles that resemble icy snowballs or frosted rocks orbiting Saturn. (b) The particles are so numerous and so far away from us. (Remember, distant galaxies look solid, too, although they are made of billions of separate stars.)

9.17 SATURN'S MOONS

Saturn has 18 confirmed and several suspected moons (Table 8.3). Others may be discovered as scientists continue analyzing the voluminous Voyager encounter data.

Titan is the biggest and the most intriguing moon (Figure 9.24). It has a substantial orange-colored atmosphere. The atmosphere is mostly nitrogen, with hydrocarbons such as methane. Prebiotic processes may be going on there. A dense haze hides Titan's surface. The moon is likely made of rock and ice with a possible sea of liquid methane and ethane. Surface temperature and pressure are 94 K (–292° F).

Mimas, Enceladus, Tethys, Dione, and Rhea are apparently mainly water ice. Except for Enceladus, all are heavily cratered. Hyperion and Iapetus also

Figure 9.24. Saturn's largest moon, Titan (without the thick atmosphere), and its medium-sized moons, shown at the same scale in this composite of Voyager images.

seem to be mostly water ice. Hyperion looks oldest, with evidence of meteoritic bombardment. Iapetus has icy and dark material on opposite sides. Phoebe's revolution is retrograde.

Irregular shapes of eight small moons indicate they are fragments of shattered larger bodies. Prometheus shepherds the inner edge and Pandora, the outer edge of the F ring. Their gravitational effects at varying distances may cause the ring's kinks.

Which of Saturn's moons is the largest and the most intriguing? Explain.

Answer: Titan. It has a substantial atmosphere of mostly nitrogen and hydrocarbons. Prebiotic processes may be going on there.

9.18 PLANET URANUS

Uranus was the first planet identified by means of a telescope. British astronomer William Herschel (1738–1822) discovered it in 1781 using a 150-mm (6-inch) telescope he made himself. Almost named for King George III, Uranus was finally named traditionally for the Greek god of the heavens.

Uranus, with a maximum magnitude of +5.7, looks like a small disk (sometimes tinted blue) through a telescope. You might spot it with your eye or binoculars if you know exactly where to look (see Useful Resources).

The planet was largely a mystery until Voyager 2 flew within 81,500 km (50,600 miles) of its cloud tops in 1986. Voyager 2 sent back 7000 images of the Uranian system.

Tipped on its side and surrounded by a system of narrow rings, Uranus resembles a giant bull's eye. The angle between its axis and the pole of its orbit is a unique 98°. The north and south polar regions are alternately exposed to sunlight and darkness as Uranus orbits the Sun. Its rotation is retrograde.

Possibly early in its history Uranus suffered a collision with a planet-sized body that knocked it over.

The atmosphere is mostly hydrogen and about 15 percent helium, with smaller amounts of methane and other hydrocarbons. It looks blue because methane preferentially absorbs red light from sunlight. The atmosphere has clouds running east to west like those of Jupiter and Saturn.

Winds blow in the same direction as the planet rotates, at speeds of 460 to 160 m/sec (90 to 360 miles per hour). Surprisingly, sunlit and dark cloud tops show the same average temperature, about 60 K (–350° F).

Voyager 2 detected haze around the sunlit south pole and large amounts of ultraviolet light, called **dayglow**, radiated from the sunlight hemisphere.

Uranus has a magnetosphere with intense radiation belts and radio emissions. Its magnetic field axis is tilted 60° to the rotational axis. The magnetic field is comparable to Earth's in intensity, but it varies much more because it is so off center. It may be generated by an electrically conductive, superpressurized ocean of water and ammonia located between the atmosphere and rocky core.

A rotating cylindrical magnetotail extends at least 10 million km (6 million miles) behind the planet. It is twisted into a long corkscrew shape by the planet's extraordinary rotation.

Eleven narrow rings are distinctly different from Jupiter's and Saturn's (Figure 9.25). They are very dark and are composed mainly of large icy chunks several feet across. Intense irradiation may have darkened any methane trapped in their icy surfaces. Transient dusty lanes appear and disappear. The chunks must bump into each other and make the fine dust that seems to be spread throughout the ring system. Atmospheric drag due to a hydrogen corona that Voyager 2 observed around Uranus may cause dust particles to spiral in to the planet.

Incomplete rings and varying opacity in several of the main rings suggest that the ring system may have formed after Uranus. Ring particles may be remnants of a moon that was broken by a high-velocity impact or torn up by gravitational effects.

Why is Uranus tipped on its axis? _____

Answer: Possibly early in its history Uranus suffered a collision that knocked it over.

9.19 URANUS'S MOONS

Five large moons and at least 10 small moons orbit Uranus (see Table 8.3).

The biggest moons look like tiny bright dots through our large telescopes. Titania was the first discovered, in 1787, and Miranda the last, in 1948. Voyager 2 found that the moons are dark gray ice-rock conglomerates, apparently made of about 50 percent water ice, 20 percent carbon and nitrogen-based materials, and 30 percent rock.

Miranda, the smallest of the five, looks strangest (Figure 9.25). It has huge fault canyons as deep as 20 km (12 miles), terraced layers, a chevron feature, large relief mountains, ridges, and rolling plains. This mixture of different terrain types on older and younger surfaces suggests diversity of tectonic activity, violent impacts, and tidal heating caused by Uranus's gravitational tug.

The two largest moons, Titania and Oberon, are about half the size of our Moon. Ariel has the brightest and possibly youngest surface, with many fault valleys and apparent extensive flows of icy material. Titania has huge fault systems and canyons that evidence past geologic activity. The surfaces of darkest Umbriel and Oberon look heavily cratered and old, indicating little past geologic activity.

The small moons were discovered by Voyager 2. Puck, the biggest, is 155 km (96 miles) in diameter. They are made of more than half rock and ice.

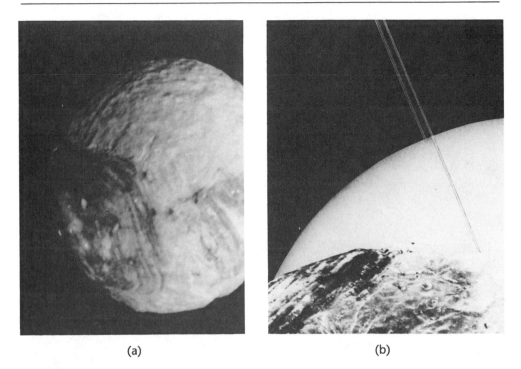

(a) (b)

Figure 9.25. (a) Uranus's moon Miranda in a mosaic of images from Voyager 2. (b) Artist's conception of the rings around Uranus overlaid on a montage of Voyager 2 images. The surface of Miranda is in the foreground.

Portia and Rosalind are shepherding the outermost ring, epsilon, apparently keeping it in a narrow region.

(a) How do the surfaces of Ariel and Umbriel differ? _____

_____

(b) What do these differences indicate? _____

_____

Answer: (a) Ariel is bright with many fissures and apparent extensive flows of icy material; Umbriel is dark, heavily cratered, and old. (b) Ariel: geologic activity; Umbriel: little geologic activity.

9.20 PLANET NEPTUNE

When Voyager 2 flew within 5000 km (3000 miles) of **Neptune** in 1989, the planet was the most distant one from the Sun. The 8000 images Voyager sent

Figure 9.26. Neptune from Voyager 2, computer enhanced to bring out small features. Bright clouds near Great Dark Spot-1989 changed their appearances in a few hours.

back gave us our first good look at the Neptunian system (Figure 9.26). Like Uranus, Neptune has a thick hydrogen, helium, and methane cloud cover that appears bright blue.

Neptune's discovery was a triumph for theoretical astronomy. Uranus did not follow the path Newton's law of gravity predicted it should. Astronomers John Adams (1819–1892) in England and Urbain Leverrier (1811–1879) in France calculated that its motion was being disturbed by another planet's gravity. They predicted where that unknown planet should be in the sky.

In 1846 astronomer Johann Galle (1822–1910) at the Berlin Observatory in Germany pointed to the predicted spot and found Neptune. The planet was named for the Roman god of the sea.

Although Neptune, the smallest planet of the gas giants, receives only 3 percent as much sunlight as Jupiter, it has a dynamic atmosphere. There the strongest winds on any planet blow westward, opposite the direction of rotation. Several large, dark spots and high, long, bright clouds, streaks, and plumes appear.

Great Dark Spot 1989 was a giant storm the size of Earth, which resembled Jupiter's Great Red Spot. It circuited Neptune every 18.3 hours. Winds blew up to 2000 km (1200 miles) per hour nearby. Great Dark Spot 1989 vanished and a new northern one, Great Dark Spot 1994, was photographed by the Hubble Space Telescope.

Neptune's magnetic field is highly tilted, 47° from the rotation axis. It may be characteristic of flows in the interior. The magnetic field causes radio emission and weak auroras.

Voyager found four rings circling Neptune. They are so diffuse and the material in them is so fine that they could not be fully resolved from Earth.

Why was Neptune's discovery a triumph for theoretical astronomy? _____

_____

_____

Answer: Theory predicted that an unseen planet must exist. Neptune was discovered by looking for it at the spot in the sky predicted by theory.

9.21 NEPTUNE'S MOONS

Neptune has eight confirmed moons (Table 8.3).

Triton is the largest and the most interesting (Figure 9.27). Voyager data

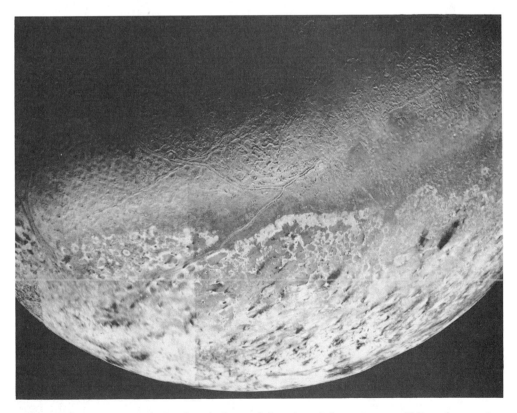

Figure 9.27.   Neptune-facing hemisphere of the planet's largest moon, Triton, in a composite of about 12 images from Voyager 2.

showed that Triton's surface has methane ice. Recent infrared measurements showed carbon monoxide and carbon dioxide and carbon dioxide ices as well. Active geyserlike eruptions shoot invisible nitrogen gas and dark dust particles up several kilometers. The surface temperature is the coldest observed in the solar system, about 38 K (–391° F). The large south polar cap is slightly pink. From the ragged edge northward Triton is darker and redder, possibly colored by ultraviolet light and magnetospheric radiation acting on its atmospheric and surface methane.

A very thin atmosphere extends some 800 km (500 miles) above Triton's surface. Surface pressure is about 14 microbars, or 1/70,000 of Earth's. Nitrogen ice particles may form thin clouds a few kilometers above the surface.

Six small, dark moons discovered by Voyager 2 remain close to Neptune's equatorial plane. They were given names from mythology's water gods. Proteus, the biggest, is 420 km (250 miles) in diameter. Like the rings, the small moons are probably fragments of larger moons shattered in collisions.

With a relatively high density and retrograde orbit, Triton does not seem to be an original member of Neptune's family. Suggest a possible origin.

Answer: Perhaps Neptune captured Triton from an originally eccentric orbit.

9.22 PLANET PLUTO

Pluto, usually the planet farthest from the Sun, is a frozen world named for the Greek god who ruled over the underworld of dead souls (Figure 9.28). It is very faint, with its brightest magnitude +14.

U.S. astronomer Clyde Tombaugh (1906–) discovered Pluto in 1930 while searching for a Planet X whose gravity is the postulated cause of irregularities in the orbits of Uranus and Neptune. Pluto may shine as brightly as it does by reflecting sunlight from frozen methane mixed with other ices that cover its surface. If so, Pluto is even smaller than it is now estimated to be. Apparently it is not massive enough to be the proposed Planet X which some astronomers still expect to discover.

Recent observations of a stellar occultation confirmed that the planet has an atmosphere of methane gas. When a star passed behind Pluto, its light was gradually dimmed and then gradually returned, rather than being suddenly blocked and unblocked.

A relatively large moon, Charon, orbits about 19,000 km (11,000 miles) above the planet (Table 8.3). Charon was first seen as a bulge on Pluto's image by U.S. astronomer James W. Christy in 1978. It is named for the mythological boatman who ferried souls of the dead to Hades. Charon orbits Pluto in

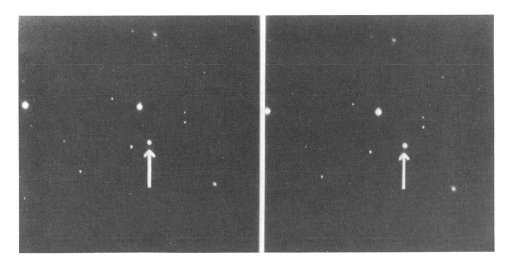

Figure 9.28. Two photographs of Pluto, showing motion of the planet in 24 hours.

exactly the time Pluto takes to rotate once. So an astronaut would always see Charon in the same spot in the sky and only from one hemisphere from Pluto.

Of all the planets, Pluto has the most eccentric orbit. Pluto moved closer to the Sun than Neptune in 1980, reached perihelion in 1989, and is now heading away from the Sun. On March 14, 1999, Pluto will cross 6.1 AU above Neptune's orbit outbound to be the outermost planet again.

Not only are Pluto's moon and orbit unique, most of its other properties are unusual, too. Its origin in the solar system is puzzling. Pluto might be a large icy planetesimal, an escaped moon of Neptune, or an interstellar object captured by the Sun's gravity as it passed too close by.

No spacecraft have been to Pluto yet, and none are being planned now.

Pluto is so far out in the solar system that it has not completed even one trip around the Sun since its discovery.

Refer to Figure 9.28. Can you explain how astronomers verified that Pluto is not a star? _____

_____

_____

_____

Answer: Photographs of Pluto taken at different times show how it "wanders" relative to the background stars. (Its discoverer Clyde Tombaugh found Pluto by painstakingly comparing millions of star images and planet suspects on pairs of photographs of sections of the sky taken on different dates.)

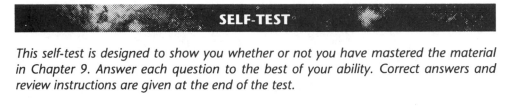

SELF-TEST

This self-test is designed to show you whether or not you have mastered the material in Chapter 9. Answer each question to the best of your ability. Correct answers and review instructions are given at the end of the test.

1. Match each planet to a famous feature visible in a small telescope.

   ____ (a) Phases.   (1) Mars.
   ____ (b) Polar ice caps.   (2) Jupiter.
   ____ (c) Great Red Spot.   (3) Saturn.
   ____ (d) Rings.   (4) Venus.

2. Match common features to correct planet pairs.

   ____ (a) Alternate, parallel, dark,   (1) Mercury and Venus.
             and light cloud bands.   (2) Jupiter and Saturn.
   ____ (b) Many craters and mountains.   (3) Uranus and Neptune.
   ____ (c) Thick hydrogen, helium, and
             methane cloud covers.

3. List three reasons why Venus would be a very unpleasant planet to visit.

   (1) _____

   (2) _____

   (3) _____

4. Figure 9.29 shows Venus, Earth, and Mars in orbit around the Sun. What letter in the diagram indicates the following points?

   (1) Venus is an evening star. _____
   (2) Venus is in a new phase. _____
   (3) Mars is at opposition. _____
   (4) Mars is not visible in our nighttime sky. _____

5. Sketch and label the three principal layers of the Earth.

   (1) _____ ; (2) _____ ; (3) _____

6. Give three observations that support the theory of plate tectonics (continental drift).

   (1)_____

   _____

   _____

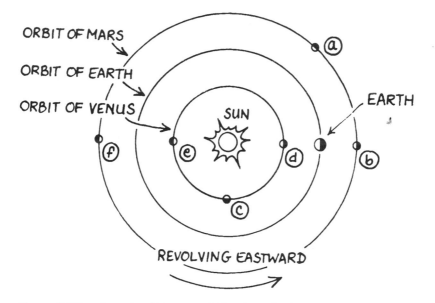

Figure 9.29. Aspects of Venus and Mars from Earth.

(2)_____

_____

(3)_____

_____

7. Describe the scene, atmosphere, and temperatures at the Viking 1 landing site on Mars. _____

_____

_____

8. State two observations that indicate water may have flowed on Mars long ago.

(1)_____

_____

(2)_____

_____

9. List the most abundant gases in the atmospheres of

(a) Earth _____

(b) Mars_____

(c) Jupiter _____

(d) Saturn _____

(e) Uranus _____

(f) Titan _____

10. Match a planet or planets to a Voyager discovery:

____ (a) Encircling ring(s).

____ (b) Moon(s) orbiting the planet.

(1) Jupiter.

(2) Saturn.

(3) Uranus.

(4) Neptune.

11. Match a planet's moon to:

____ (a) Largest moon in solar system.

____ (b) Only moon known to have a substantial atmosphere.

____ (c) Most geologically active moon, with active volcanoes.

____ (d) Strangest, with mixture of young and old surfaces.

____ (e) Coldest surface, with active geyserlike eruptions.

(1) Ganymede/Jupiter.

(2) Io/Jupiter.

(3) Miranda/Uranus.

(4) Titan/Saturn.

(5) Triton/Neptune.

ANSWERS

Compare your answers to the questions on the self-test with the answers given below. If all of your answers are correct, you are ready to go on to the next chapter. If you missed any questions, review the sections indicated in parentheses following the answer. If you missed several questions, you should probably reread the entire chapter carefully.

1. (a) 4; (b) 1; (c) 2; (d) 3. (Sections 9.2, 9.9, 9.13, 9.14, 9.16, 9.18, 9.20)

2. (a) 2; (b) 1; (c) 3. (Sections 9.1, 9.3, 9.13, 9.14, 9.16)

3. Poisonous carbon dioxide atmosphere; much too hot (up to 900° F); and a crushing atmospheric pressure over 90 atmospheres. (Section 9.3)

4. (1) c; (2) d; (3) b; (4) f. (Sections 9.2, 9.9)

5. Figure 9.10: (1) crust; (2) mantle; (3) core. (Section 9.5)

6. (1) Similar plant and animal fossils are found along coastlines of South America and West Africa. (2) These coastlines seem to fit together. (3) No rocks from the bottom of the Atlantic Ocean near the coastlines are older than about 150 millions years. (Section 9.6)

7. The surface looks like a red, dry, rock-strewn desert. The sky is pink. The temperature is cold. (Section 9.10)

8. (1) Deep, winding channels that look as if they were carved by great rivers; (2) water frozen in the polar ice caps. (Section 9.11)

9. (a) nitrogen (about 78 percent) and oxygen (about 21 percent); (b) carbon dioxide; (c) hydrogen and helium; (d) hydrogen and helium; (e) hydrogen and helium, with some methane; (f) nitrogen. (Sections 9.8, 9.11, 9.14, 9.16 through 9.18)

10. (a) 1, 2, 3, 4; (b) 1, 2, 3, 4. (Sections 9.15, 9.17, 9.19)

11. (a) 1; (b) 4; (c) 2; (d) 3; (e) 5. (Sections 9.15, 9.17, 9.19, 9.21)

10

THE MOON

*The stars about the lovely Moon hide their shining forms
when it lights up the Earth at its fullest.*

Sappho (c. 612 B.C.)

Fragment 4

Objectives

☆ Explain the Moon's appearance and apparent motions in the sky.
☆ Compare the Moon and Earth in diameter, mass, average density, and surface gravity.
☆ Describe the general surface features of the Moon.
☆ Compare and contrast the Moon and Earth with respect to geological activity and erosion of surface features.
☆ Outline a hypothesis of the origin of the Moon that is consistent with observations.
☆ Explain the probable origin of lunar craters and maria.
☆ Describe surface conditions on the Moon at the Apollo landing sites.
☆ Give the current model of the Moon's internal structure.
☆ List some questions about the Moon that remain to be answered.
☆ Describe the relative positions of Earth, Moon, and Sun during a solar eclipse and a lunar eclipse.

10.1 NEAREST NEIGHBOR

Poets have always been enchanted by the beautiful Moon. At magnitude –12.5, a full Moon is almost 25,000 times more brilliant than first-magnitude stars (Figure 10.1).

People once believed that the brilliant Moon influenced personal behavior directly. They practiced special rituals at full Moon. Some of the ancient

Figure 10.1. The full Moon. The six sites where U.S. Apollo astronauts landed are marked 11 through 17. Prominent maria, craters, and mountain ranges are named on your Moon Map.

names for a Moon goddess were Diana, Lunae, Selene, and Cynthia. Words such as "Moonstruck" and "lunacy" originally referred to a madness that changed with the phases of the Moon.

Today we know more about the Moon than about any other neighbor in space. It is the closest sky object of all, located an average of 384,400 km (240,000 miles) from Earth. Robot spacecraft and astronauts have been to the Moon and returned thousands of photographs, scientific data, and surface samples.

Six U.S. **Apollo Moon missions** (1969–1972) landed men there with cameras and scientific experiments and returned 380 kg (837 pounds) of Moon rock for laboratory study. The Apollo instruments sent back data until 1977 when they were finally turned off for budgetary reasons.

The Moon shines by reflecting sunlight. Its average visual **albedo**, the proportion of incident sunlight that the Moon reflects back into space, is only 11 percent. Most of the sunlight that shines onto the airless Moon's surface is absorbed.

Why do you think the full Moon is the brightest light in the night sky?

_____

_____

Answer: Because it is so much closer to Earth than all other sky objects.

10.2 IN THE SKY ☽

If you look at the Moon regularly, you will observe its two apparent motions in the sky in addition to its phases (Section 8.4).

You will see the Moon rise in the east, move westward across the sky, and set every day, because the Earth rotates daily.

You will also observe that the Moon changes its location with respect to the stars about 13° to the east every day, because the Moon moves with respect to the Sun daily as the Earth-Moon system revolves around the Sun every year (Figure 10.2).

Explain why the Moon rises an average of about 50 minutes later each day than it did the day before. _____

_____

_____

_____

Answer: At Moonrise the Moon is located in a particular constellation. About 24 hours later, when Earth has turned completely around, those same stars rise again, but meanwhile the Moon has moved about 13° eastward with respect to the stars and so does not rise until later.

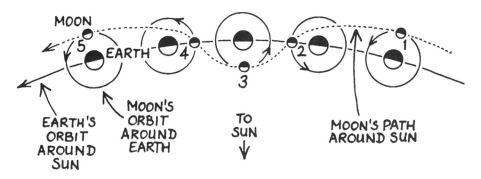

Figure 10.2. The Earth-Moon system's revolution around the Sun. The waviness of the Moon's orbit is greatly exaggerated for clarity.

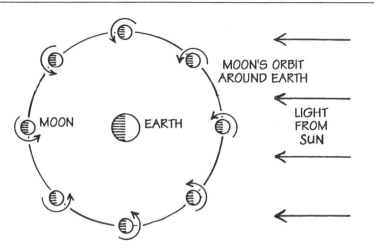

Figure 10.3. The Moon's synchronous rotation. Note: The same side of the Moon always faces Earth.

10.3 ROTATION

Earth's gravity has locked the Moon into **synchronous rotation**. The Moon rotates on its axis every 27.3 days, the same amount of time it takes to travel around Earth. That makes the same side of the Moon face Earth at all times (Figure 10.3).

Notice that you observe the same features of the "man in the Moon" all month long, but never see the back of his head. (The visible disk of the Moon appears to shift, called **libration**, due to slight variations in the Moon's motions, so you can actually see a total of 59 percent of its surface over time.)

The Moon's rotation period and its revolution period are probably not equal by coincidence, but are most likely equal because of eons of tidal friction.

Why were people able to see only one side of the Moon before spacecraft flew to the far side? _____

_____

Answer: The Moon's rotation period is equal to its period of revolution around Earth, so the same side of the Moon always faces Earth.

10.4 SPECIAL EFFECTS

You may observe other dramatic changes in the Moon's appearance.

A **lunar halo**, or ring around the Moon, is really not near the Moon at all.

Ice crystals high up in Earth's atmosphere refract Moonlight as it passes through, creating the halo effect.

When the Moon is near the horizon, it may look red. From that position, moonlight travels a longer path through the atmosphere to our eyes than when the Moon is overhead. Moonlight (reflected sunlight) consists of all visible colors. Short (blue) moonlight rays are scattered out and the long (red) rays, which penetrate the atmosphere more readily, color the Moon red.

A full Moon looks bigger when it is near the horizon than when it is high in the sky. The size of the Moon is always the same. No one knows exactly what causes this common **Moon illusion.** Perhaps comparing the Moon with local objects as opposed to distant stars makes the difference psychologically.

The **harvest Moon** is the full Moon nearest the time of the autumnal equinox. It rises earlier in the evening than usual, lighting up the sky to give farmers extra hours for harvesting. Harvest Moon occurs when the angle between the ecliptic and the horizon is near minimum.

Do you think the Apollo astronauts saw a "ring around the Earth" from the Moon's surface? Explain._____

_____

_____

Answer: No. The ring around the Moon is an illusion caused by ice particles in Earth's atmosphere. The Moon has no atmosphere or water to create the illusion of a ring around Earth in the sky.

10.5 SIZE

The Moon is unusually large for a satellite, in comparison to its parent planet. The Moon's size can be found from measurements of its angular diameter and its distance from Earth.

The distance to the Moon has been measured to a fantastic accuracy of one part in 10 billion (a few centimeters) by timing how long it takes a laser light beam to reach reflectors there and return.

The equatorial diameter of the Moon is 3476 km (2160 miles). The equatorial diameter of Earth is 12,756 km (almost 8000 miles).

Compare the size of the Moon to that of Earth. _____

Answer: The diameter of the Moon is about 1/4 that of Earth.

Solution: Diameter of Moon/diameter of Earth $\cong$ 3500 km/13,000 km $\cong$
        (2000 miles/8000 miles) = 1/4.

10.6 DENSITY

The Moon's mass, measured from the accelerations the Moon produces on spacecraft, is 7.35×10^{22} kg, or 1/81 that of Earth.

The Moon's average density is 3.34 t/m³, or roughly 3/5 that of Earth.

The Moon's surface gravity is only about 1/6 that of Earth because of its small mass and size. That means an 84-kg (180-pound) astronaut would weigh only 14 kg (30 pounds) on the Moon's surface.

Suggest a reason why the Moon's average density is less than Earth's.

_____

_____

_____

Answer: The Moon must be made almost entirely of silicate rocks like those of Earth's crust and mantle, and be poor in iron and other metals. (Moon rocks analyzed so far are made of the same chemical elements as Earth rocks, but the proportions are different.)

10.7 DATA

Review the properties of the Moon you have learned so far by completing Table 10.1. Review Sections 8.4 and 8.10, if necessary.

TABLE 10.1   Properties of the Moon

Quantity	Value
(a) Average distance from Earth	_____
(b) Diameter	_____
(c) Sidereal orbital period (fixed stars)	_____
(d) Synodic orbital period (phases)	_____
(e) Rotation period	_____
(f) Mass	_____
(g) Average density	_____
(h) Surface gravity	_____
(i) Albedo	_____
(j) Apparent magnitude of full Moon	_____
(k) Average velocity in orbit	_____

Answer: (a) 384,400 km (240,000 miles); (b) 3476 km (2160 miles), or 1/4 that of Earth; (c) 27.3 days (27.321 66 days); (d) 29.5 days (29.530 59 days); (e) 27.3 days; (f) 7.35 × 10^{22} kg, or 1/81 that of Earth; (g) 3.34 t/m^3, or 3/5 that of Earth; (h) 1/6 that of Earth; (i) 0.11; (j) –12.5; (k) 1.02 km/sec (2295 miles per hour).

10.8 OBSERVING

The Moon has long been a favorite target for binoculars and low-power telescopes because it is close enough to see in great detail.

The Moon Map at the back of this book has been drawn especially to make it easy for you to identify prominent surface features. It shows the Moon as it appears around its highest in the sky in the northern hemisphere. Compass directions on your Moon Map correspond to sky directions. This map is oriented with north at the top, the way we see the Moon with our eyes or through binoculars.

(Through many telescopes the Moon appears inverted, with north at the bottom. For astronauts on the Moon or topographic maps of the surface, east and west are interchanged as with Earth maps; north and south remain unchanged.)

When Galileo first pointed his telescope at the Moon, he mistakenly thought the large, relatively smooth dark areas he saw were oceans. He called them **maria** (singular **mare**), meaning "seas."

The Moon missions found no water on the Moon. The maria are actually dry lava beds. They contain **basalt**, a type of rock produced by the cooling of molten lava from volcanoes. Mare Imbrium, the Sea of Showers, the largest mare on the Moon's visible side, is about 1100 km (1700 miles) across.

The brighter areas of the Moon are called **highlands**. They are higher, more rugged, older regions than the maria. Highlands contain light-colored igneous rocks. They cover about 80 percent of the surface.

What are the maria that form the features of the "man in the Moon?"

Answer: Solidified lava beds.

10.9 CRATERS

The Moon is pitted with **craters**, or holes in the surface.

The craters are customarily named after famous scientists and philosophers such as Copernicus and Plato. The largest, flat-floored craters, such as Clavius, nearly 240 km (150 miles) across, are called **walled plains**. The smallest are known as **craterlets**.

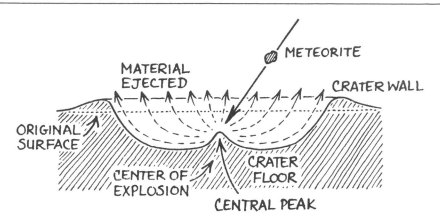

Figure 10.4. Formation of a typical impact crater.

Typical craters are circular, ranging in size from tiny pits to huge circular basins hundreds of meters across with walls ranging up to 3000 m (10,000 feet) in height. Most were probably blasted out by high-speed meteorites crashing onto the Moon (Figure 10.4).

Figure 10.5. The far side of the Moon was first photographed by Russian robot Luna 3 on October 4, 1959.

The heat of impact vaporizes a meteorite and some of the ground it penetrates. The hot, vaporized material expands violently and explodes, forming a circular crater with a high rim and often a central peak. Ejected material falls back around the crater, often forming smaller secondary craters.

Bright **rays** radiate for hundreds of kilometers across the surface from young prominent craters. These are apparently splash patterns made of ejected material from the impact explosion.

The best time to observe craters and mountains is when they are near the sunrise or sunset line, called the **terminator**. Then the Sun's low elevation above the ground causes shadows that highlight surface relief.

The sunrise terminator moves from right to left over the Moon's surface between its new and full phases. The sunset terminator does the same between full and new Moon. When the Moon is full, the maria stand out prominently, but the lack of surface shadows makes the surface relief hard to discern.

Spacecraft photographs show that the far side of the Moon has craters and highlands, but it does not have large maria, which are so conspicuous on the near side. The cause of the observed differences between the near and far sides of the Moon is not yet understood (Figure 10.5).

What probably produced most craters on the Moon? _____

Answer: Meteorites crashing into the surface.

10.10 EXPLORING

When U.S. astronaut Neil Armstrong first set foot on the Moon for "all mankind" on July 20, 1969, he entered a strange, desolate world.

The entire surface of the Moon is covered by powdery soil called **regolith**, which is produced by the shattering of surface rock during prolonged meteorite bombardment. Many areas have a lot of loose rock ranging in size from pebbles to huge boulders.

No water flows, nothing grows. No water, fossils, or organisms of any kind have been found in laboratory analysis of Moon rocks and soil. This lack of evidence of life suggests that the Moon is and has always been lifeless.

No blue sky, white clouds, or weather of any kind appear above the Moon because the Moon has no appreciable atmosphere. Ghostly silence prevails in the absence of air to carry sounds.

Days and nights are long—14 Earth-days each. The surface temperature ranges from about 120° C (250° F) when the Sun is at its highest point to about –130° C (–280° F) at night.

Apollo astronauts proved that the Moon is accessible for human activity (Figure 10.6). In the twenty-first century, astronauts at a Moon base could perform astronomical and other scientific research and also process resources such as oxygen and metals to support new space exploration.

Figure 10.6.   Twelve U.S. Apollo astronauts spent a total of 300 hours on the Moon. The last three pairs drove lunar rovers and explored the surface around their landing sites.

Why would it be useful to place a large optical telescope on the Moon?

_____

_____

Answer: In the absence of air or weather on the Moon, the seeing (Section 2.21) would always be good.

10.11 BOMBARDMENT

The major agents of erosion of the Moon's surface are **micrometeorites**, tiny grains of rock and metal, that crash into the Moon at speeds of up to 113,000

Figure 10.7. Historic Apollo 11 Moon visit. Edwin E. Aldrin photographed by Neil Armstrong, whose reflection was captured in the helmet visor.

km (70,000 miles) an hour. Large meteorites also collide with the Moon occasionally.

Micrometeorites are about 10,000 times less effective in changing the Moon's surface than are air and water on Earth. They remove only about 1 millimeter of lunar surface in a million years.

Explain why Neil Armstrong's first boot print on the Moon will probably look the same millions of years from now as it did in 1969 (Figure 10.7).

_____

_____

Answer: Erosion on the Moon is due primarily to micrometeorite bombardment and happens much more slowly than erosion by air and water on Earth.

10.12 MOUNTAINS

Mountains on the Moon are named after the great ranges on Earth, such as the Alps. They are different from ours in both chemical composition and appearance because they were produced and shaped by different forces.

The highest rugged mountain peaks on the Moon tower over 8000 m (29,000 feet), as does Mount Everest, Earth's highest mountain.

What two major factors that constantly change the shape of Earth's mountains do not shape Moon mountains? Explain. _____

_____

Answer: Water and atmosphere. No mountain streams pour down these ranges. No atmospheric storms ever rage to wear away the surface.

10.13 HISTORY

Lunar scientists have reconstructed this life story of the Moon from U.S. Apollo and Russian Luna Moon-flight data.

The oldest Moon rocks, collected in the highlands, are about 4.3 billion years old. A few tiny green rock fragments are about 4.6 billion years old. The youngest, from the maria, were formed about 3.1 billion years ago.

The manner and place of birth of the Moon remain a mystery.

Moon rocks are richer in silicates and poorer in metals and volatile elements than Earth rocks. So the Moon is probably not a part of Earth that was torn off. It probably did not form by the accretion of many smaller particles in the solar nebula, either.

The popular **impact-ejection hypothesis** says that soon after the Earth formed, a planet-sized body crashed into it. The impact ejected a giant glob of material from the Earth, which broadened to a ring around our planet. Material in the ring then collected to form the Moon.

During its first billion years of life, the young Moon was heavily bombarded by meteorites of all sizes. They produced great craters and melted the surface, now the lunar crust.

When the Moon was about 1 billion years old, the interior was greatly heated by radioactive elements. Volcanoes poured huge floods of hot basaltic lava over the surface and into the craters. This molten lava solidified, forming the maria.

About 3 billion years ago the Moon cooled off significantly, and volcanic activity largely stopped. Except for minor lava flows and a relatively small number of large impact craters like the young (about 1 billion years old) Copernicus, the Moon has not changed since. Seismographs left on the Moon by Apollo astronauts detected a very low level of Moonquakes.

The Moon's airless, dry, stable surface preserves a historical record of ancient bombardments that must have been common to all terrestrial planets.

How does the story of the Moon's geological activity differ from Earth's?

_____

_____

_____

Answer: The Moon became essentially dead geologically after the first 2 billion years of its life in contrast to Earth, which is still very much alive with volcanism, mountain building, and drifting continents.

10.14 INSIDE THE MOON

Geologists draw their current picture of the Moon's interior from spaceflight data. Gravity field measurements revealed **mascons**, mass concentrations,

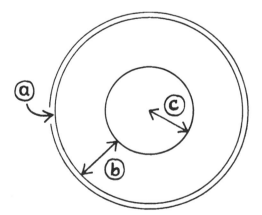

Figure 10.8.   The Moon's structure.

submerged in the circular maria. The existence of mascons plus the absence of major Moonquakes suggest that the Moon has a cold, thick, rigid outer layer, or crust. The crust is about 60 km (40 miles) thick on the near side and thicker on the far side.

Beneath the crust, extending down about 1000 km (600 miles), is the mantle. The physical characteristics of the core, extending the last 700 km (400 miles) to the center, are still unknown. The core may be partly molten, at a temperature up to 1500 K.

The Moon does not have a magnetic field today, but old lunar rocks indicate that it once did.

Refer to Figure 10.8. Identify the crust, mantle, and core, and indicate the approximate depth of each layer. (a) _____ ;

_____

(b)_____ ;

(c) _____

Answer: (a) Crust: 60 km (40 miles) on the near side and thicker on the far side; (b) mantle: 1000 km (600 miles); (c) core: 700 km (400 miles).

10.15 SURFACE CONDITIONS

Questions about the Moon still abound. Future research and exploration of the surface, as well as analysis of the rest of the lunar material brought back from Apollo and Luna missions, is expected.

Summarize what we have learned about the surface of the Moon so far.

_____

_____

Answer: Your paragraph should describe the maria, craters, mountain ranges, absence of air and water, length of day and night, and surface temperatures.

10.16 ECLIPSE OF THE SUN

A **solar eclipse** occurs when the Earth, new Moon, and Sun are directly in line (Figure 10.9).

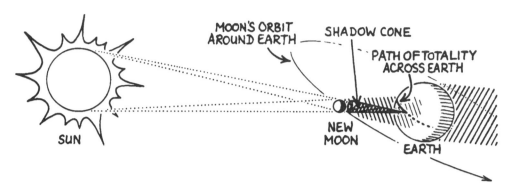

Figure 10.9. Solar eclipse (not to scale).

The eclipse is **total** when the Moon is closer to Earth than the length of its shadow cone. The Moon looks bigger than the Sun and blocks the Sun's bright disk from view.

Totality lasts only a few minutes and can be seen only at successive places along a narrow curved path (a few hundred kilometers wide) inside the Moon's shadow on Earth. The maximum duration of totality is 7.5 minutes, which is not predicted to occur again until July 16, 2186.

Over a wider region bordering both sides of the path of totality a **partial eclipse** is seen. A partial eclipse may also occur when the Moon is not quite close enough to the Sun-Earth line to block all of the Sun from view (Figure 10.10).

An **annular eclipse** occurs when the Moon is farther from Earth than the length of its shadow cone. The Moon looks smaller than the Sun, and blocks from view all of the Sun's bright disk except for an outer ring (annulus) of sunlight (Figure 10.11).

It is thrilling to observe a total eclipse of the Sun! When the Moon passes in front of the bright Sun, an unnatural darkness spreads across the sky, the temperature drops, and stars and planets shine in the daytime sky. People

Figure 10.10. Partial solar eclipse.

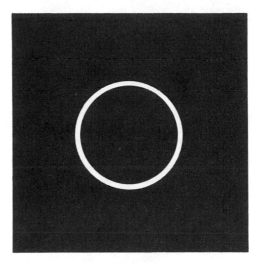

Figure 10.11. Annular solar eclipse.

once connected the Sun's disappearance with terrifying events. Today professional and amateur astronomers eagerly go around the world to observe a total solar eclipse and gather astronomical data

Your chances of observing a total solar eclipse from your hometown are very small, since the occurrence at any one location on Earth averages only about 1 in 360 years. You might consider joining an eclipse expedition to view this spectacular natural event. The most important coming total solar eclipses are listed in Table 10.2.

What phase must the Moon be in for an eclipse of the Sun to occur?_____

Answer: New.

TABLE 10.2   Total Solar Eclipses

Date	Duration of Totality (minutes)	Where Visible
1994 November 3	4:23	South America
1995 October 24	2:1	South Asia
1997 March 9	2:5	Siberia, Arctic
1998 February 26	4:09	Central America
1999 August 11	2:23	Central Europe, Central Asia
2001 June 21	4:57	South Atlantic, Africa
2002 December 4	2:04	Southern Africa, Australia
2003 November 23–24	1:57	Antarctica

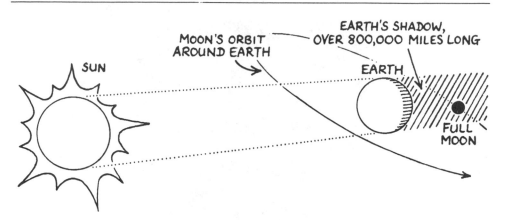

Figure 10.12. Lunar eclipse (not to scale).

10.17 ECLIPSE OF THE MOON

A **lunar eclipse** occurs when the Sun, Earth, and full Moon are directly in line (Figure 10.12).

The Moon darkens when it enters Earth's shadow, but it still gets some sunlight that is refracted around Earth by our atmosphere. Clouds, dust, and pollution affect the color and brightness of the Moon's appearance, usually making it dull red.

More than 2000 years ago, the Greeks observed that during a lunar eclipse, Earth's shadow appears circular on the Moon. Philosopher Aristotle (384–322 B.C.) cited this evidence to support the theory that the Earth is a sphere rather than flat. Astronomer Eratosthenes (c. 276–194 B.C.) made the first fairly accurate measurement of the diameter of the Earth.

TABLE 10.3   Total Lunar Eclipses

Date		Duration of Totality (minutes)	Visible from North America
1993	June 4	96	No
1993	November 29	46	Yes
1996	April 4	86	Yes
1996	September 27	70	Yes
1997	September 16	62	No
2000	January 21	76	Yes
2000	July 16	107	No
2001	January 9	60	No
2003	May 16	52	Yes
2003	November 8–9	22	Yes
2004	May 4	76	No
2004	October 28	80	Yes

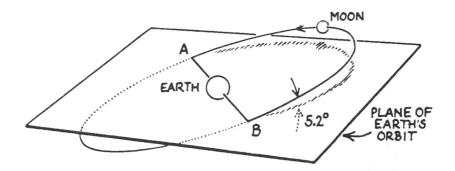

Figure 10.13. Unfavorable conditions for an eclipse.

Your chances of seeing a total lunar eclipse are much greater than your chances of seeing a total solar eclipse (Table 10.3). Lunar eclipses, when they occur, can be seen from any place on Earth where the Moon is shining. They last much longer than solar eclipses. The maximum duration of totality is 1 hour, 47 minutes, predicted to occur next on July 16, 2000.

What phase must the Moon be in for a lunar eclipse to occur? _____

Answer: Full.

10.18 ECLIPSE TIMES

The maximum number of solar and lunar eclipses that can occur in one year is seven.

Eclipses do not occur every time we have a new or full Moon, as you might expect. The Moon's orbit is tilted 5.2° to the plane of Earth's orbit. Most months the Moon is above or below the Sun-Earth line at new and full phase, so no eclipse can occur (Figure 10.13).

The Moon's orbit crosses the plane of Earth's orbit at two points called **nodes**. The nodes move slowly westward, called **regression of the nodes**, because of the Sun's gravitational pull.

Figure 10.14.   The plane of the Moon's orbit is tilted 5.2° to the plane of Earth's orbit.

Examine Figure 10.14. Explain why an eclipse can occur only when the Moon is at point A or point B. _____

_____

Answer: Then the Sun, Earth, and Moon are directly in line.

10.19 OCCULTATIONS

Occultation means the eclipse of one sky object by another.

An occultation by the Moon is the most frequent and easiest type for you to observe. The Moon often passes between Earth and a star or planet, causing it to suddenly disappear and to reappear after the Moon moves by. Predictions of lunar occultations are published in current astronomical publications (see Useful References).

Jupiter is over 40 times bigger than the Moon. How is it possible for the Moon to occult, or hide, Jupiter? _____

_____

Answer: Jupiter is much farther away than the Moon, so it appears much smaller.

SELF-TEST

This self-test is designed to show you whether or not you have mastered the material in Chapter 10. Answer each question to the best of your ability. Correct answers and review instructions are given at the end of the test.

1. Why do earthbound observers always see the same side of the Moon?

   _____

   _____

2. The Moon is about what fraction of Earth in (a) diameter? _____

   (b) mass? _____ (c) average density? _____

   (d) surface gravity? _____

3. Match the lunar features with their names:

   ____ (a) Dry lava beds.   (1) Craters.
   ____ (b) Holes in the surface.   (2) Highlands.
   ____ (c) Light-colored, higher, rugged   (3) Maria.
             older regions.   (4) Mascons.
   ____ (d) Submerged clumps of mass.

4. Suppose you were leading an expedition to explore the surface of the Moon. Which of the following would be useful? (a) extra oxygen tanks; (b) flare pistol and flares; (c) flashlight; (d) magnetic compass; (e) matches; (f) star chart; (g) umbrella; (h) watch _____

   Explain. _____

   _____

5. What is the probable origin of most craters on the Moon? _____

   _____

6. Why do the Moon's surface features change so much more slowly than Earth's? _____

   _____

   _____

   _____

7. How old are the (a) oldest and (b) youngest Moon rocks that have been col-lected? (a) _____ ;

   (b) _____

8. Sketch the Moon's interior; identify its three main layers. (a) _____ ;

   (b) _____ ; (c) _____

9. List three questions about the Moon that remain to be answered.

   (1)_____

   _____

   (2)_____

   _____

   (3)_____

   _____

10. What must the phase of the Moon be for (a) the occurrence of an eclipse of the Sun? _____ (b) an eclipse of the Moon? _____

ANSWERS

Compare your answers to the questions on the self-test with the answers given below. If all of your answers are correct, you are ready to go on to the next chapter. If you missed any questions, review the sections indicated in parentheses following the answer. If you missed several questions, you should probably reread the entire chapter carefully.

1. The Moon's rotation period and its period of revolution around Earth are equal, called synchronous rotation.   (Section 10.3)

2. (a) 1/4 diameter;   (b) 1/81 mass;   (c) 3/5 density;   (d) 1/6 surface gravity.   (Sections 10.5 through 10.7)

3. (a) 3;   (b) 1;   (c) 2;   (d) 4.   (Sections 10.8, 10.9, 10.14)

4. (a);   (c);   (f);   (h). Because the Moon has no air, water, or magnetic field, nothing that requires any of these would be useful.   (Sections 10.8, 10.10, 10.14)

5. Meteorites crashing into the surface.   (Section 10.9)

6. There is no air or water to cause erosion on the Moon as they do on Earth. There is no comparable geological activity, either. Micrometeorites crashing into the Moon are the major agents of erosion on the Moon's surface. (Sections 10.11 through 10.13)

7. (a) About 4.6 billion years old;   (b) 3.1 billion years old.   (Section 10.13)

8. Refer to Figure 10.8. (a) Crust;   (b) mantle;   (c) core.   (Section 10.14, Figure 10.8)

9. What is the chemical composition of the surface at places away from the few Apollo and Luna landing sites? Are water or other volatiles, perhaps brought by meteorites or comets, frozen in the polar region? How did the Moon originate? (You may have thought of others.)   (Sections 10.6, 10.13 through 10.15)

10. (a) New;   (b) full.   (Sections 10.16, 10.17)

11

COMETS, METEORS, AND METEORITES

When beggars die, there are no comets seen:
The heavens themselves blaze forth the death of princes.

William Shakespeare (1564–1616)
Julius Caesar, Act II, ii:30

Objectives

☆ State why comets and meteorites are of interest to scientists.
☆ Describe the current theory of the origin and composition of comets.
☆ Explain in terms of the current model of comet structure the changes in a comet's appearance as its distance from the Sun changes.
☆ Specify the relationship between comets and meteor showers.
☆ Distinguish between a meteoroid, meteor, and meteorite.
☆ Give the composition and probable origin of meteorites.
☆ List some possible effects on Earth of a major comet or meteorite impact.

11.1 COMETS

Bright comets have always fascinated people (Figure 11.1). Unlike ordinary stars, these fiery-looking objects appear and disappear unpredictably. Records of bright comets go back to the fourth century B.C. Throughout history people have dreaded them as omens of human disasters such as wars or famines.

Today we know that a **comet** is an icy member of our solar system. Comets travel in elliptical orbits around the Sun and follow the basic laws of physics. They are not supernatural signs at all.

Figure 11.1. Famous Halley's Comet on March 16, 1986, five weeks after its most recent perihelion passage.

(a) What was the common historical view of comets? _____

_____

(b) What is the modern astronomical view of comets? _____

_____

Answer: (a) Comets were considered supernatural signs foretelling human misfortune. (b) Comets are icy members of the solar system that follow natural physical laws and have no hidden meaning.

11.2 SIGNIFICANCE OF COMETS

Comets that appear in our sky are important, even if they don't shine brilliantly. They are probably the only objects left that are made out of the original material from which the whole solar system formed about 5 billion years ago. Earth, Moon, and other celestial bodies have all been changed by tectonic processes, erosion, or numerous collisions. Only comets remain basically as they were in the beginning.

Comet Halley is the most studied comet so far, and astronomers assume that all comets have a similar makeup. Researchers in 50 countries deployed their most sophisticated instruments on Earth and in space in an International Halley Watch during the 1986 apparition. They propose to send future spacecraft to rendezvous with another comet and to scrutinize it close up for several years.

Why are comets important? _____

_____

_____

Answer: They are our best source for observing the original material out of which everything in the solar system was formed.

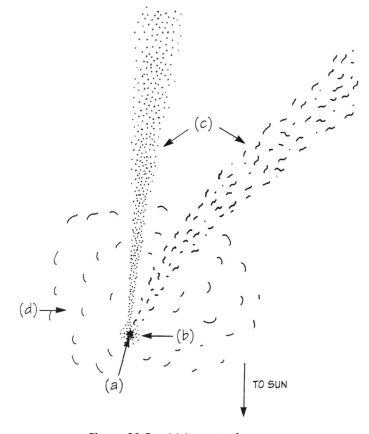

Figure 11.2. Main parts of a comet.

11.3 COMET STRUCTURE

Comets were named for their appearance. Both the Greek word ***kometes*** and the Latin word ***cometa*** mean "long-haired."

When it shines in the sky, a bright comet has a head with a starlike core called the **nucleus** surrounded by a glowing halo called the **coma** and long transparent **tails**. The nucleus is several kilometers in size. The coma may extend 100,000 km (60,000 miles) or more outside the nucleus. Tails can stream millions of kilometers into space.

Ultraviolet observations from spacecraft reveal a huge enveloping **hydrogen cloud**. It can grow up to tens of millions of kilometers wide. This cloud is not visible from Earth.

Refer to Figure 11.2. Identify the main parts of a typical bright comet.

(a) _____ ; (b) _____ ; (c) _____ ; (d) _____

Answer: (a) Nucleus; (b) coma; (c) tails; (d) hydrogen cloud.

11.4 THE NUCLEUS

Billions of comets probably orbit far out in the solar system, but you can't see them from Earth. They shine in the sky only when they travel near the Sun. The most widely accepted description of a typical comet is the **dirty snowball model**, proposed by U.S. astronomer Fred Whipple in 1950 (Figure 11.3).

When a comet is far out in the solar system it consists of only a nucleus. Its shape and surface are irregular. The nucleus is made of mostly water ice and other frozen gases (the "snow") mixed with stony or metallic solids (the "dirt"). It has very low density and surface gravity.

In an historic first, the European Space Agency (ESA) robot spacecraft

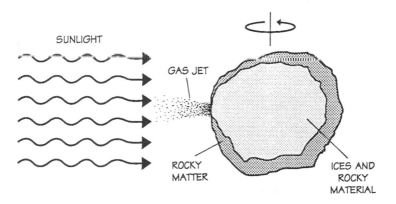

Figure 11.3.   Dirty snowball comet model.

Giotto plunged through Comet Halley's head and penetrated to within 600 km (375 miles) of the icy spinning nucleus. Giotto sampled dust and gas directly, and radioed data to Earth for analysis.

The nucleus was found to be dark black, roughly 8 km by 15 km (5 miles by 9 miles) in size and rotating about once every 2.2 days.

Surface irregularities included cracks, crevasses, and possible craters. About 10 percent of the surface had rifts that spewed jets of dust and gas, primarily water vapor, toward the Sun. The rest had a sooty black insulating dust layer, estimated to be about a centimeter thick, that probably was left behind when many passages of the comet around the Sun drove off volatile gases.

Gases detected coming off of the nucleus were 80 percent water vapor by volume and other compounds, including carbon dioxide, carbon monoxide, ammonia, and methane. Some dust grains seemed to be silicates, while others contained virtually only the elements carbon, hydrogen, oxygen, and nitrogen.

Scientists are intrigued that carbon exists in complex organic molecules in Comet Halley, which could have biological significance.

What is the nucleus of a comet made of? _____

_____

_____

Answer: The dirty snowball model describes a comet nucleus as mostly water ice and other frozen gases mixed with solids.

11.5 THE COMA

As a comet nucleus comes in from the edge of the solar system to within a few hundred million kilometers of the Sun, it heats up. Gases sublime and escape to space with dust from its surface. The comet's gravity is too weak to hold back the gases and dust. They expand outward around the nucleus for thousands of kilometers, forming the coma.

The comet shines because the gases fluoresce and the dust reflects sunlight. Astronomers use large telescopes to image about 25 of these blurry blobs of light each year.

What causes the coma to develop?_____

_____

Answer: The Sun's heat (which causes sublimation and expansion of gas and dust particles).

11.6 THE TAILS

When a comet moves near the Sun, it may develop tails of gases and dust released from the nucleus.

Ultraviolet radiation tears the gases apart into free radicals (molecular fragments) and ions. Ions interact with the charged particles blowing out from the Sun in the solar wind. The ions are ultimately swept millions of kilometers straight back into a **gas**, or **ion, tail**.

Radiation pressure, or intense sunlight striking, pushes the dust particles outward. The comet keeps moving, and a **dust tail** curves behind it. Comet tails are so thin you can see right through them to the stars on the other side.

Neutral molecules and atoms continue to expand outward from the nucleus until they are ionized. The most common atom, hydrogen, forms the huge hydrogen cloud. The hydrogen cloud surrounding the nucleus of Comet Halley grew to several hundred thousand kilometers in diameter.

Effects of hydrogen ions released by Comet Halley on the solar wind were detected as far as 35 million km (21 million miles) from the nucleus. **A bow shock**, where cometary gases block and slow the solar wind, was found some 400,000 km (240,000 miles) in front of the comet.

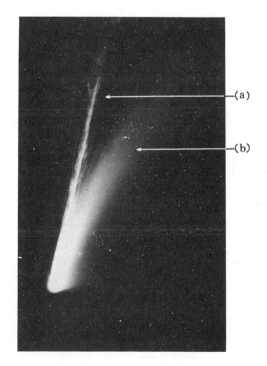

Figure 11.4.   Comet Mrkos with typical tails.

Refer to Figure 11.4. Identify the gas or ion tail and the dust tail and state the cause of each. (a) _____ ;

(b) _____

Answer: (a) Ion tail; solar wind; (b) dust tail; radiation pressure.

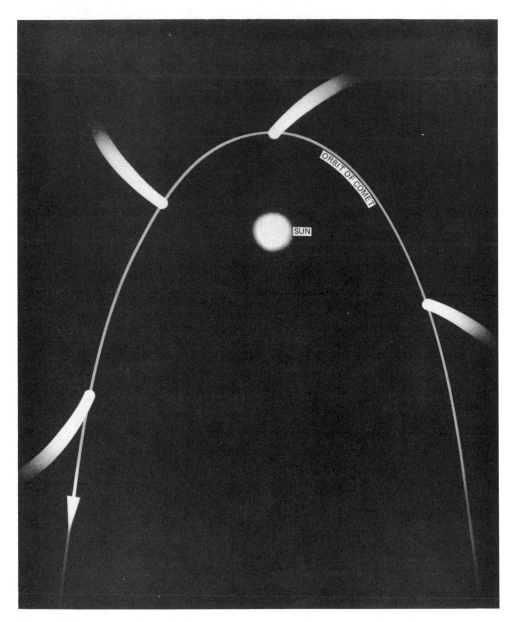

Figure 11.5.   A comet's perihelion passage.

11.7 DISAPPEARANCE

As the comet accelerates inexorably nearer the Sun, its fate is unpredictable. Powerful jets of gas and dust from the nucleus may change its orbital motion.

If a comet rounds the Sun whole, it continues along its orbit back to frigid outer space. Some cometary material is left behind and the rest refreezes. Coma and tails vanish.

Some comets pass so close to the fiery Sun that they shatter or disintegrate. Occasionally one crashes directly into the Sun and disappears.

Refer to Figure 11.5. Why do comets go back to outer space tail first?

_____

_____

Answer: Comet tails are caused by solar radiation pressure and solar wind, which are always directed away from the Sun, so the tail must also point away from the Sun.

11.8 ORIGIN OF COMETS

The comets we observe originate in a vast shell of icy objects located 50,000 to 100,000 times farther from the Sun than Earth is, according to a widely

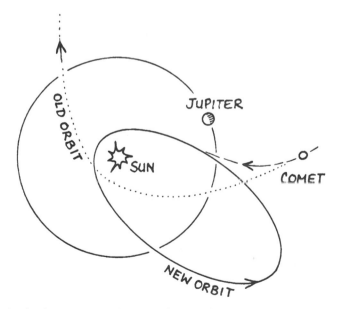

Figure 11.6. Jupiter's strong gravity perturbs a passing long-period comet into a new, short-period orbit around the Sun.

accepted model developed in the 1950s by Dutch astronomer Jan Oort (1900–1992). That **Oort cloud**, about a third of the way out to neighbor stars, may hold 100 billion incipient comets.

Occasionally a passing star tugs on a comet, slows it down in its motion, and sends it plummeting toward the Sun. That comet will be a **long-period comet**, with an almost parabolic orbit and a period of revolution around the Sun of 200 hundred to millions of years.

If the comet passes near a giant planet, notably Jupiter, it will be affected by that planet's strong gravity. Then the comet may crash into the planet, or speed up and head out of the solar system, or move into a relatively short-period elliptical orbit around the Sun (Figure 11.6).

Where do the comets we observe probably originate? _____

Answer: In a vast cloud of comets near the edge of the solar system.

11.9 PERIODIC COMETS

Astronomers have catalogued about 150 **short-period, or periodic comets,** that have periods of revolution around the Sun of a few years or decades up to 200 years. They shine periodically in the sky every time they come close to the Sun.

The most consistently bright and most famous is Comet Halley, with 30 consecutive perihelion passages recorded since 240 B.C. Sighted telescopically for over three years before and after its February 9, 1986 perihelion passage, Comet Halley is also the best-analyzed comet so far.

Table 11.1 lists some comets that have appeared several times in our sky. What is the shortest known period of revolution of a comet? _____

Answer: 3.3 years (Comet Encke).

Table 11.1   Some Periodic Comets

Comet	Period[a] (years)	Closest Approach to Sun (in AU)
Encke	3.3	0.33
Temple	5.3	1.37
Kopff	6.5	1.59
Wolf	8.4	2.50
Halley	76.1	0.59

[a]Period may change over time.

11.10 COMET FATE

A periodic comet cannot be reactivated to grow a new coma and tails indefinitely. Its nucleus loses a surface layer a few meters deep each time it rounds the Sun. Dust and gas litter its orbit. Comet Halley leaves about 1 percent of its mass behind during each perihelion passage (Figure 11.7).

Finally, a periodic comet will lose all its volatile material. Large chunks and numerous small fragments of nonvolatile solids may survive. The debris continues to orbit around the Sun like tiny planets.

Summarize five changes of appearance that a comet undergoes as it travels in its orbit around the Sun. _____

_____

_____

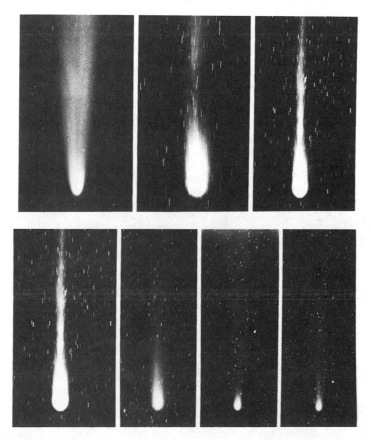

Figure 11.7. Comet Halley on seven different days as it receded from the Sun after its apparition in 1910.

Answer: 1. Far from the Sun, a comet consists of a nucleus of frozen gases and dust. 2. Coma forms as a comet approaches the Sun. 3. Close to the Sun, tails form. 4. After going around the Sun, much cometary material refreezes. 5. Far from the Sun again, coma and tails are gone.

11.11 COMET HUNT

Several new comets are discovered every year. Professionals find some among their data in observatories, and amateurs discover the others.

Comets are usually named for their discoverers. Exceptions, such as Halley and Encke, honor the people who first computed their orbits. The first three people to report seeing a new comet may have their names permanently attached to it. Since hunting comets is a popular international activity, we sometimes get real tongue twisters, such as the short-period (5.3 years) Comet Honda-Mrkos-Pajdusakova!

How could a comet give you immortality? _____

Answer: Discover one, and it will carry your name.

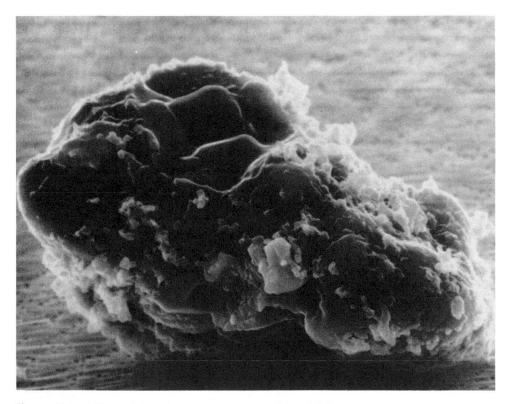

Figure 11.8.   Piece of dust from a comet, magnified 15,000 times.

11.12 INTERPLANETARY REMAINS

Countless bits and pieces of matter, called **meteoroids**, populate the inner solar system.

Earth is surrounded by interplanetary dust. It is observed at infrared wavelengths. Meteoroids enter Earth's atmosphere continually. Astronomers collect them at high altitudes, from arctic ice sheets and from the ocean floor for analysis in laboratories. These meteoroids are similar to dust grains ejected from the nucleus of Comet Halley (Figure 11.8).

What is a meteoroid?_____

Answer: A solid particle orbiting the Sun in space.

11.13 SHOOTING STARS

Did you ever make a wish on a "shooting star" or "falling star"? These light flashes are not stars at all. They are **meteors**, streaks of light created by meteoroids that are plunging through Earth's atmosphere at speeds up to 72 km (45 miles) per second. Air friction burns these tiny particles when they are 60 to 110 km (40 to 70 miles) above Earth.

On any clear, dark night you can see an average of six meteors an hour flashing unpredictably in the sky. Meteors occur but are not visible during the daytime because the sky is too bright.

A large meteoroid creates an exceptionally brilliant meteor, called a **fireball**. The largest may partially survive their fiery plunge. On March 8, 1976, a spectacular red fireball the size of a full Moon was seen by tens of thousands of people over a wide area in northeast China. A violent break-up occurred when the fireball was 17 km over Kirin City. After its explosive impact with the ground, large and small fragments were turned over to scientists for study.

What is a "shooting star" or "falling star"? _____

Answer: A "shooting star" or "falling star," or meteor, is the streak of light that can be seen when a meteoroid is burned upon entry into Earth's atmosphere.

11.14 METEOR SHOWERS

On several predictable dates every year you may see meteors pour down from one part of the sky. This type of display is called a **meteor shower**. Meteor showers are associated with comets. They occur when Earth, moving along its

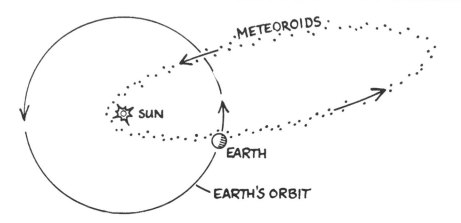

Figure 11.9. A meteor shower occurs when Earth passes near the orbit of a comet and encounters a swarm of meteoroids.

orbit around the Sun, crosses a swarm of meteoroids left behind by an active comet (Figure 11.9).

In 1910, when Earth was about to travel right through the tail of Halley's Comet, people panicked. What would you expect to happen if Earth passed through a comet's tail? _____

Answer: A bright (but harmless) meteor shower.

11.15 BEST METEOR DISPLAYS

During a meteor shower, all the meteors appear to come from one common point in the sky, called the **radiant** of the shower. Meteor showers are usually named for the constellation where they seem to originate, such as the Perseids from Perseus and the Orionids from Orion.

You can usually see more meteors after midnight than before, because Earth, moving along its orbit, is then traveling head-on into the swarms of particles. Meteor showers are best seen with your unaided eye on nights when the Moon is not bright. A full Moon can blot out a meteor shower.

Table 11.2 lists famous annual meteor showers. Since shower activity can vary with time, it is best to consult a current astronomical publication (see Useful Resources) for details of this year's best showers.

Refer to Table 11.2. What is the name and date of maximum of the largest summer meteor shower observable from around 40° N latitude? _____

Answer: Perseids; August 12.

TABLE 11.2 Principal Annual Meteor Showers

Shower	Date of Maximum	Approximate Hourly Rate	Associated Comet
Quadrantids	January 3	30	
Lyrids	April 23	8	1861I
Eta Aquarids	May 4	10	possibly Halley
Delta Aquarids	July 30	15	
Perseids	August 12	40	1962III
Orionids	October 21	15	possibly Halley
Taurids	November 4	8	Encke
Leonids	November 16	6	1866I Temp
Geminids	December 13	50	
Ursids	December 22	12	Tuttle

11.16 ROCKY LANDINGS

When a piece of stone or metal from outer space lands on Earth, it is called a **meteorite**.

Within recent history, no case has been documented of anyone being killed by a stone from the sky. Probably hundreds of tons of cosmic material fall through the atmosphere every year, but only about two or three meteorites in a decade land in places where people live, and even more rarely is any injury reported.

The largest meteorite ever found, the Hoba West, weighs an estimated 66 tons. It still lies in South-West Africa where it landed. A meteorite is usually named after the post office closest to its landing site. Many large meteorites are on display in museums worldwide (Table 11.3).

What is a meteorite? _____

Answer: A piece of stone or metal from outer space.

TABLE 11.3 Large Meteorites on Display in the U.S.

Meteorite	Approximate Weight	Present Location
Ahnighito (Greenland)	34 tons	American Museum of Natural History (New York City)
Willamette (Oregon)	14 tons	American Museum of Natural History
Furnas County (Nebraska)	1 ton	University of New Mexico
Paragould (Arkansas)	800 pounds	Chicago Natural History Museum

11.17 MAKEUP OF METEORITES

If you need extra spending money, find a meteorite! Scientists and collectors have paid generously for genuine material from outer space. Meteorites are the only material from space, besides the Moon rocks and soil from the Apollo and Luna trips, that scientists can study close up.

Meteorites are divided into three main types. (1) **Iron meteorites** are about eight times as dense as water and consist mostly of iron (about 90 percent) and nickel. (2) **Stony-iron meteorites** are about six times as dense as water. They contain iron, nickel, and silicates. (3) **Stony meteorites** are about three times as dense as water. They have a high silicate content, and only about 10 percent of their mass is iron and nickel.

Iron meteorites are found most often. Stony meteorites look like ordinary Earth rocks. They are not usually noticed unless they are seen falling. Laboratory analysis confirms their extraterrestrial origin. Table 11.4 lists the occurrence of the different types of iron and stony meteorites.

Most meteorites are probably fragments of asteroids shattered by collisions, since these objects have similar compositions. They appear to be about 4.6 billion years old, which is the approximate age of the whole solar system.

When a stony meteorite with a high carbon and some water content, called a **carbonaceous chondrite**, is found, it adds excitement to the search for life in space (Chapter 12). These finds show that prebiotic matter formed beyond Earth!

The 4.5-billion-year-old stony Murchison Meteorite fell in Victoria, Australia, in 1969. It contained simple amino acids, which make up proteins, and nucleic-acid bases that carry and replicate genetic information. It also had organic chemicals similar to lipids (structural components of living cells).

The Allende Meteorite fell in northern Mexico, also in 1969. It is one of the largest recorded carbonaceous chondrite falls. Almost 2 tons of some of the most primitive material in the solar system were recovered.

A collection of more than a thousand uncontaminated meteorites recovered from the ice in Antarctica include some carbonaceous chondrites containing amino acids.

Some "Moon" meteorites are close in composition to rocks collected on the Moon by the Apollo astronauts. A few "Mars" meteorites have gases trapped inside that are nearly identical chemically to the atmosphere of Mars. Possibly a comet or asteroid struck Mars violently, causing chunks of rock to escape Mars's gravity and orbit through space, ultimately to be captured by Earth's gravity.

TABLE 11.4   The Occurrence of Meteorite Types

Types	Meteorites Seen Falling	Meteorite Finds
Irons	6%	66%
Stony-irons	2%	8%
Stones	92%	26%

Why are meteorites important to scientists? _____

_____

_____

Answer: They are primitive extraterrestrial material that scientists can examine closely to determine more about the solar system.

11.18 IMPACTS ON EARTH

You may wonder what would happen if a comet or large meteorite struck Earth.

Large meteorites carve out huge craters in planets and moons when they crash. Earth must have been hit often by meteorites early in its history. Ancient craters are erased by geologic activity and erosion. Major impacts are now extremely rare. You can view the most recent, Meteor Crater near Winslow, Arizona, U.S. (Figure 11.10). It was created by a meteorite impact over 25,000 years ago.

A comet nucleus could impact with the energy of millions of hydrogen

Figure 11.10.   Meteor Crater in Arizona, U.S., is about 1.5 km across and 180 m deep.

bombs. But most astronomers figure the chances of a direct hit on Earth are extremely remote, and the impact was more likely a meteorite. Most comets never come anywhere near Earth in their travels around the Sun.

On June 8, 1908 in Siberia a mysterious explosion with the force of about 12 megatons about 8 km up in the air flattened over 1000 square km of forest near the Tunguska River. It killed reindeer 25 km (40 miles) away. A large meteorite or comet could have blown up and caused the damage.

A major impact by a space object could have caused the catastrophic extinction of dinosaurs and many other plant and animal species about 65 million years ago. Researchers found enriched deposits of iridium in the K-T boundary, the geological layer of sediments laid down at the end of the Cretaceous Era and the start of the Tertiary Era 65 million years ago. Iridium is more abundant in comets, meteorites, and asteroids than in Earth's crust. Shock-melted mineral spherules and soot were also found in and near the K-T boundary. A huge crater buried under the Yucatan Peninsula in Mexico is the postulated impact site.

Is it likely that Earth will be hit by a comet nucleus or large meteorite in the near future? _____

Answer: No.

SELF-TEST

This self-test is designed to show you whether or not you have mastered the material in Chapter 11. Answer each question to the best of your ability. Correct answers and review instructions are given at the end of the test.

1. Why do modern astronomers use their most sophisticated instruments to study comets? _____

   _____

   _____

2. What is a comet nucleus made of? _____

   _____

   _____

3. State two important discoveries that were made about the nucleus of Comet Halley made during its 1986 perihelion passage. (1) _____

   _____

   _____

   (2)_____

   _____

   _____

4. Describe five changes of appearance that a periodic comet undergoes as it travels in its orbit around the Sun. _____

   _____

   _____

   _____

   _____

5. Sketch and identify the main parts of a typical bright comet.

   (a) _____ ; (b) _____ ;

   (c) _____ ; (d) _____

6. Describe the origin and extinction of periodic comets. _____

_____

_____

_____

_____

7. Match each description with the correct item.

    ____ (a) "Shooting star" or "falling star."   (1) Meteor.

    ____ (b) Small particle orbiting the Sun.   (2) Meteorite.

    ____ (c) Solid body that reaches Earth.   (3) Meteoroid.

8. Explain the relation between comets and meteor showers.

_____

_____

_____

9. Describe the composition and probable origin of meteorites.

_____

_____

_____

_____

10. List the following in order of increasing distances from the Sun: asteroid belt, Earth, Oort comet cloud, Pluto. _____

_____

11. Explain why meteorites are of interest to scientists. _____

_____

_____

ANSWERS

Compare your answers to the questions on the self-test with the answers given below. If all of your answers are correct, you are ready to go on to the next chapter. If you missed any questions, review the sections indicated in parentheses following the answer. If you missed several questions, you should probably reread the entire chapter carefully.

1. Comets are considered to be the most unchanged objects left that are made out of the original material from which the whole solar system formed. (Section 11.2)

2. A comet nucleus is made of mostly water ice and other frozen gases mixed with solids according to the dirty snowball model.   (Section 11.4)

3. (1) The nucleus is dark black and potato-shaped, roughly 15 km (9 miles) long. (2) The surface has cracks, crevasses, and possible craters, a sooty black insulating dust layer, and jets of dust and gas escape near perihelion. (Section 11.4)

4. Far from the Sun, a comet consists of a nucleus of frozen gases and dust. Coma forms as a comet approaches the Sun. Tails form close to the Sun. After going around the Sun, a comet refreezes. Far away from the Sun, a comet consists of a nucleus again.   (Sections 11.3 through 11.7, 11.9, 11.10)

5. Refer to Figure 11.2. (a) Nucleus;   (b) coma;   (c) tail;   (d) hydrogen cloud.   (Section 11.3)

6. Periodic comets probably originate in the vast Oort comet cloud near the edge of the solar system. Jupiter's strong gravity redirects those that pass nearby from long-period to short-period orbits around the Sun. After numerous perihelion passages, periodic comets finally lose all their volatile material. Only fragments of nonvolatile solids may survive.   (Sections 11.8 through 11.10)

7. (a) 1;   (b) 3;   (c) 2.   (Sections 11.12, 11.13, 11.16)

8. Meteor showers occur when Earth, moving along its orbit around the Sun, crosses a swarm of meteoroids left behind in space by a comet.   (Section 11.14)

9. Iron meteorites—mostly iron (about 90 percent) and nickel; stony irons—iron, nickel, and silicates; stony meteorites—high silicate content, only about 10 percent iron and nickel by mass; probable origin: asteroid belt. (Section 11.17)

10. Earth, asteroid belt, Pluto, Oort comet cloud.   (Sections 11.8, 11.17)

11. Because they originate in space and can help us understand the history and composition of our own planet, Earth, and of the solar system.   (Sections 11.16, 11.17)

12

LIFE ON OTHER WORLDS?

We are attempting to survive our time so we may live into yours. We hope someday, having solved the problems we face, to join a community of galactic civilizations. This record represents our hope and determination, and our good will in a vast and awesome universe.

President Jimmy Carter, 1977
Record attached to Voyager spacecraft

Objectives

☆ Describe the molecular basis of Earth life.
☆ Give the evidence that indicates life may have evolved spontaneously from nonliving molecules on Earth.
☆ Explain a scientific theory of the origin and evolution of intelligent life on Earth.
☆ Describe the search for life on Mars.
☆ State the evidence for the existence of planetary systems other than our own.
☆ List the factors involved in estimates of the statistical chances for extraterrestrial intelligent life.
☆ Describe past and present human research and exploration in space.
☆ State the dominant current scientific view of interstellar voyages and UFOs.
☆ Describe several projects in which scientists have searched for or are planning to search for extraterrestrial intelligence.

12.1 PROMISE

No one knows whether **extraterrestrial life**, life beyond Earth, exists. Life on Earth could be a unique cosmic accident. But humans have compelling clues that we are not alone.

Biochemists find that all living organisms on Earth depend on a few basic organic molecules, or molecules containing carbon, which they can manufacture from gas atoms in the laboratory.

Astronomers detect these basic atoms and molecules of life in our solar system, stars, and interstellar dust clouds. They have also found meteorites with amino acids and lipidlike chemicals.

Physicists assume that the natural laws which rule physical and chemical events on Earth apply everywhere in the universe.

If life on Earth evolved from nonliving molecules in a series of physical and chemical processes, then life may have developed elsewhere among the more than 200 billion stars in our Milky Way Galaxy or in one of the 100 billion other galaxies in the universe.

The search has begun!

Two U.S. robot Viking spacecraft performed the first tests for life on another planet, Mars, in 1976. Radio astronomers are currently probing the sky for intelligent signals from other civilizations.

Explain why many scientists think extraterrestrial life may exist. _____

_____

_____

_____

Answer: The basic molecules of life have been found in space and manufactured in the laboratory. If living organisms evolve from nonliving molecules in a series of physical and chemical reactions and are not a unique cosmic accident, then life may have occurred on other worlds.

12.2 COSMIC ORIGINS

The most basic characteristics that separate a living organism from a nonliving one are the ability to reproduce and metabolism. What causes the spark of life? No one knows, but a **cosmic evolution theory** ties the appearance of living organisms to universal forces in the following way:

Our universe exploded into existence in the Big Bang between 10 and 20 billion years ago. The first elements were hydrogen and helium. Subsequently the universe expanded and cooled. Galaxies and stars formed. Heavier elements were slowly produced by nucleosynthesis inside massive stars. Supernovas sprayed enriched material back into space where it was recycled.

About 5 billion years ago, our Sun condensed from an enriched interstellar cloud with biotic elements and dust grains. Earth and the other solar system bodies took shape in a collapsing, cooling disk of material orbiting the infant Sun.

At first Earth's surface was tumultuous and fiery. Active volcanoes spewed hot lava and gases continually. Meteorites and comets crashed, adding more biotic elements to the primitive Earth.

During the next billion years Earth cooled. Outgassing produced an atmosphere and ocean.

Scientists have energized a mixture of compounds of hydrogen, carbon, oxygen, and nitrogen gases like those in Earth's earliest atmosphere and have produced organic molecules, including amino acids, the basic molecules of life. The Sun's ultraviolet rays, cosmic rays, lighting flashes, and shock waves from geological activity were all available energy sources to bind the atmospheric gases into more complex organic molecules 4 billion years ago.

Another possibility is that thermal vents on the ocean floor were the cradle of life, since the conditions there seem to be conducive to the formation of organic molecules.

Gradually, organic molecules accumulated in Earth's seas. As their concentration increased, collisions between molecules in the water joined small molecules into larger ones. Water was important in this process, because it speeds chemical reactions by facilitating collisions between molecules.

Perhaps a billion years passed as more and more complex molecules formed. Eventually RNA (ribonucleic acid) and DNA (deoxyribonucleic acid) molecules were formed that carried the genetic codes for replication. The threshold from nonliving to living matter was crossed.

Why is water important in the chemical evolution of the basic organic molecules of life?_____

_____

Answer: Water speeds up chemical reactions by enabling molecules to collide with each other.

12.3 EVIDENCE

The common virus suggests that living organisms can evolve out of nonliving molecules, because it has characteristics of both.

A virus, too small to be seen other than under an electron microscope, is essentially a strand of DNA or RNA. A virus cannot provide its own energy or replicate itself outside of living cells. It survives when the cells it infects supply the energy necessary for growth and the means of reproduction.

If the sharp division between living and nonliving matter is artificial and viruses actually lie on a continuum, they would be somewhere near the middle. Perhaps a similar, ancient protocell was the precursor of life on Earth.

What evidence exists that living organisms may evolve out of nonliving molecules? _____

Answer: The virus has characteristics of both living organisms and nonliving molecules.

12.4 EVOLUTION

The principle of natural selection, or the survival of the fittest, asserts that all living creatures on Earth evolved from simple one-celled organisms.

Microfossils in rocks over 3 billion years old are evidence that life existed on Earth as simple one-celled plants called algae and organisms such as bacteria when those rocks formed (Figure 12.1).

The first living species reproduced. But offspring are never exact copies of their parents. Some variations in characteristics are always introduced whenever reproduction takes place.

Those individuals with favorable variations that helped them to survive the environment were the fittest. They had the best chance of reaching adulthood and reproducing themselves. Thus favorable traits were passed on, while unfavorable traits died out, by natural selection. Slowly over a long period of time a new species could have formed from the original one.

Multicelled organisms, which appeared about a billion years ago, and sexual reproduction speeded up evolutionary diversification.

The fossil record of the last 600 million years indicates that on many occasions there were mass extinctions of species followed by the arrival of numerous new, diverse species. The first fish were in the ocean about 425 million years ago. Reptiles appeared about 325 million years ago. After the extinction of the dinosaurs 65 million years ago, small mammals multiplied. Finally, some 40,000 years ago humans with cognitive intelligence arrived.

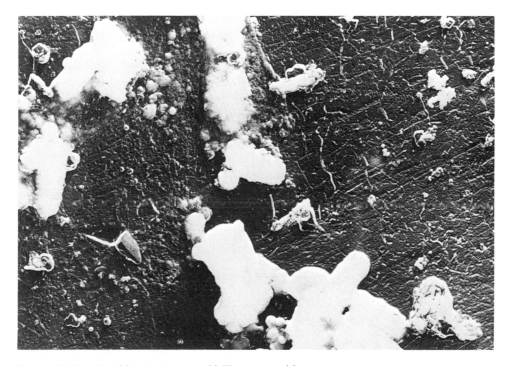

Figure 12.1. Fossil bacteria, several billion years old.

In this way, over billions of years under the varying environmental conditions that existed on Earth, all living things, including modern humans, may have evolved from simple cells.

Suggest an environmental pressure that could produce a critical evolutionary advance. _____

Answer: A drastic world-wide change in climate. (You may have thought of others.)

12.5 NEARBY PLANETS

Life may also have evolved on a neighboring planet. The Sun's **habitable zone**(ecosphere), or region where life might most comfortably exist, is roughly between the orbits of Venus and Mars.

Venus does not seem suitable for life, because it has a prohibitively hot surface temperature up to 480° C (900° F) and is too dry.

Mars seems more suitable. It looks as though large amounts of water, essential for life on Earth, once flowed on Mars. Photographs show branching channels that look like familiar river beds and tributaries, where mighty rivers may once have run (Figure 9.18). Water exists today in the Martian polar ice caps, frost, fog, and filmy clouds.

Experiments on Earth have shown that some plants and microbes can survive environmental conditions similar to those on Mars today. If ever life evolved there, perhaps it still exists.

The U.S. Viking lander experiments were designed to detect carbon-based microbes living in the Martian soil. The results are inconclusive. Puzzling activity was detected. It might have been due to living organisms, or more likely to some unusual chemical characteristic of the Martian soil.

Jupiter and Titan may have simple microbes. Their clouds contain the gases out of which life probably evolved on Earth. There may be seas of liquid hydrogen on Jupiter or liquid nitrogen on Titan where some forms of life could evolve.

Scientists have not reached a consensus about the presence or absence of life on Mars, Jupiter, or Titan. We must wait for planet probes of the future to find out.

Give three observations that suggest life may have developed from nonliving molecules on Mars. _____

_____

_____

Answer: (1) Evidence indicates that a lot of water once flowed on the planet. (2) Mars is inside the Sun's habitable zone. (3) Some plants and microbes can survive Martian conditions.

12.6 THE ODDS

We may be the only intelligent (as we proudly call ourselves) civilization in the entire universe. Or many others may exist.

Our Sun is just one of over 200 billion stars in our Milky Way Galaxy. Roughly a billion galaxies are within sight of our largest telescopes. Many of the stars in these galaxies may have life-supporting planets circling them. And some of these planets may bear intelligent civilizations.

A way to estimate the number of intelligent civilizations in our own Milky Way Galaxy—the only ones we could hope to communicate with at present— was suggested by astronomers Carl Sagan and Frank Drake of the U.S. and I. S. Shklovsky (1916–1985) of Russia. First, estimate with some confidence (1) the number of stars in the Galaxy, (2) the fraction of those stars that have planets circling them, and (3) the average number of planets suitable for life.

With much less confidence, estimate (4) the fraction of suitable planets on which life actually developed, (5) the fraction of those life starts that evolved to intelligent organisms, and (6) the fraction of intelligent species that have attempted communication.

Then, just guess (7) the average lifetime of an intelligent civilization.

When all these factors are considered, estimates range anywhere from one intelligent civilization (ours) to a million in the Milky Way Galaxy today.

Why do you think (7), the average lifetime of an intelligent civilization, is the most uncertain number of all?_____

_____

_____

_____

_____

Answer: No one knows what happens when a civilization like ours, if we are typical, reaches a stage of technical development where it can communicate with other civilizations in our Galaxy. Will it last long enough for a conversation, or will it self-destruct with nuclear weapons, pollution, or overpopulation?

12.7 EXTRASOLAR PLANETARY SYSTEMS

The nebular theory of the formation of stars asserts that many stars must be orbited by planets (Section 4.3).

Circumstellar disks, swarms of gases and particles orbiting stars, were first observed at infrared wavelengths in 1983 (Figure 12.2). Direct photographic confirmation came first for nearby star Beta Pictoris, some 50 light-years away. Possibly the thicker disks around young stars are planetary sys-

Figure 12.2. Vega in Lyra with the first circumstellar disk ever observed, from U.S./British/Dutch Infrared Astronomical Satellite (IRAS).

tems in their early formative stages, and the thinner disks around older stars are leftover material from planets that have already formed.

Planets belonging to other stars, if they exist, cannot be seen even with our largest telescopes because they are lost in the glare of their star. Astronomers search for unseen companions using three indirect techniques:

1. Observing the proper motion of the candidate visible star.

Barnard's star in Ophiuchus has the largest known proper motion. It is the third nearest star to the Sun and has been photographed since 1937. A tiny 0".001 wobble in the proper motion has been reported. If real, the wobble could be caused by the gravitational tug of two massive planets orbiting Barnard's star. It is uncertain because the wobble is only 0.01 as big as the star's image and very difficult to detect.

Astronomers propose to measure the proper motions of at least 100 nearby stars for wobbles due to unseen planets.

2. Observing the radial velocity of the candidate visible star.

The nearby star Gamma Cephei was monitored for several years and a slight Doppler shift in its spectral lines was reported. The Doppler shift indicates radial velocity variations that suggest an orbiting planet of about 1.6 Jupiter masses.

Astronomers propose to examine the spectra of 100 candidate nearby stars for a Doppler shift in their wavelengths due to unseen planets.

3. Observing the frequency of radio pulses from millisecond pulsars (the time between successive pulses is thousandths of a second).

Recently one of the closest millisecond pulsars, PSR1257 + 12, which is 1500 light-years away in the constellation Virgo, was monitored by the Arecibo radio telescope in Puerto Rico. Small, periodic changes were detected in the frequency of radio waves received. These changes could be due to the gravitational pull on the pulsar by two or three orbiting planets.

This thesis is controversial because it requires that an orderly planetary system exist around a pulsar that was created amid an explosion's powerful radiations and chaotic conditions (Section 5.16). Proponents are looking for other perturbations due to the mutual gravitational interactions between the postulated planets themselves.

What does a wobble in the proper motion of a visible star suggest?

Answer: The star's wobble suggests the presence of unseen companions.

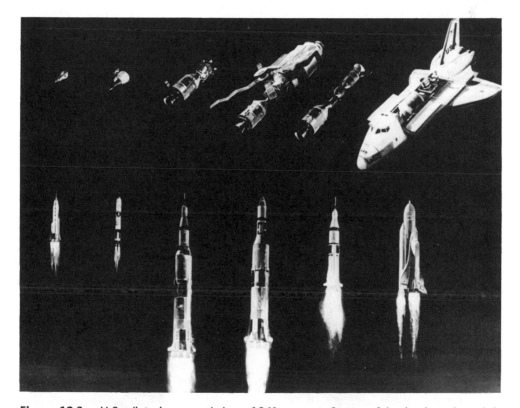

Figure 12.3.   U.S. piloted space missions, 1961–present. Spacecraft/rocket launchers, left to right: Mercury/Atlas; Gemini/Titan 2; Apollo/Saturn 5; Skylab/Saturn 5; Apollo/Saturn 1B–Soyuz; Space Shuttle Orbiter/External fuel tank and solid rocket boosters.

12.8 SPACE TRAVEL

Interstellar travel, trips to other stars, would be the most dramatic way to see if other civilizations exist. But we are not yet ready for a deep space voyage.

Even the closest stars are several light-years away, and none of our space-craft can travel anywhere near the speed of light. A trip to our Sun's nearest bright neighbor, Alpha Centauri, at the speed the Apollo astronauts traveled to the Moon, would require thousands of years.

Russian cosmonauts were the first humans in space. Yuri Gagarin (1934–1968) orbited Earth once in Vostok 1 and landed after 1 hour, 48 minutes on April 12, 1961. Valentina V. Tereshkova, the first female, circled Earth 48 times June 16–19, 1963.

For the next 20 years Russia and the U.S. developed increasingly sophisticated piloted spacecraft. Each flew just once, and most are on display in space museums today (Figure 12.3).

Experiments are now performed aboard **space stations**, Earth-orbiting satellites worked by rotating crews, and **space shuttles**, reusable craft whose missions last about a week in Earth orbit. Modular laboratories, designed for microgravity experiments, fit in space shuttle cargo bays (Figure 12.4).

Current research is focusing on the biomedical effects of microgravity and on ways to help people adapt to it, as well as on the exploration and utilization of space.

Russian cosmonauts recorded the longest piloted flights aboard a space station, Mir. Dr. Valery Polyakov set the world record of 438 days, 18 hours in 1995.

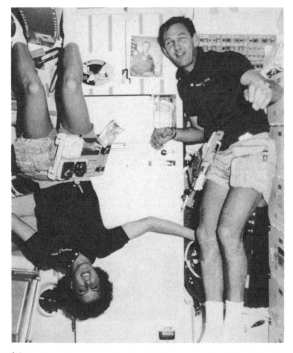

Figure 12.4. Mealtime in microgravity for U.S. space shuttle Atlantis astronauts Ellen S. Baker and Michael J. McCulley.

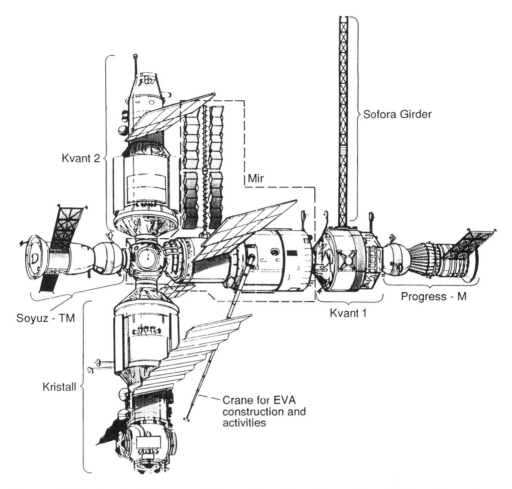

Figure 12.5. Russian Mir space station, International Space Year 1992 configuration. Modules are used for living quarters, scientific experiments, and materials processing work. Spacecraft Soyuz TM transports crew and robot Progress M, supplies. Additional modules and design alterations increase efficiency.

Cosmonauts' physiology, body chemistry, and mental health were measurably changed by long periods of weightlessness.

When did the first human go into space? _____

Answer: 1961 (Yuri Gagarin, April 12, 1961).

12.9 STAR PROBES

At present, to send high-speed robot probes to other stars would be prohibitively expensive.

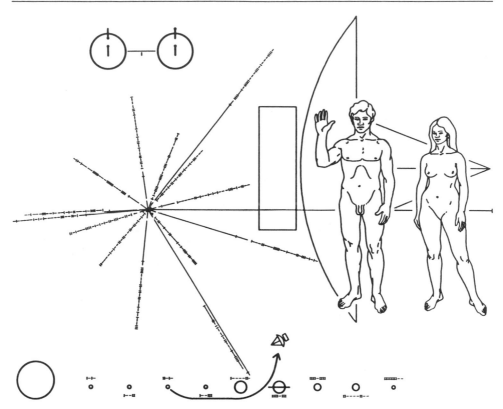

Figure 12.6. First message from Earth, the plaque aboard Pioneer 10 and 11, indicates when, where, and by whom each spacecraft was launched.

Four U.S. spacecraft are roaming into interstellar space after having completed missions to the giant planets. They carry symbolic messages for any intelligent beings they may encounter far beyond our solar system.

Pioneer 10, the first spacecraft to penetrate the asteroid belt and to provide closeups of Jupiter in 1973, was the first to leave our solar system, in 1983. Its twin, Pioneer 11, followed in 1990. Pioneer 10 and 11 each bear a symbolic plaque which is intended to show when, where, and by whom they were launched (Figure 12.6).

Voyager 1 and 2 (Section 8.12) carry a unique record with electronically encoded information, sounds, and pictures of the best of Earth and its inhabitants. A cartridge and playing instructions are included. Possible civilizations in space can hear recordings of wind and waves, birds and animals, music, a kiss and a baby's cry, and greetings in 60 languages.

Both Voyagers are programmed to study ultraviolet sources among the stars. Their fields and particles instruments are searching for the **heliopause**, where the Sun's influence ends and interstellar space begins. The Voyagers should return valuable data until the year 2015 when their nuclear power sources will no longer be able to supply necessary electrical energy.

People sometimes wonder whether Earth is visited by creatures from other worlds when they hear accounts of unidentified flying objects (UFOs).

Most scientists think aliens are the least likely explanation for UFO sightings. They ask for concrete evidence, such as a piece of an alien spacecraft, to examine in a laboratory. None has been found so far.

Why is it unlikely that Earth will be invaded by hostile beings from space, as some alarmists have suggested? _____

_____

_____

Answer: Travel across the enormous distances between stars would take much too long and exhaust too many resources to be justifiable for most purposes, if other civilizations are at our stage of development.

12.10 COMMUNICATION

We have the capability of communicating at the speed of light with other civilizations using radio waves.

With sending and receiving devices already in use, we can radio a message from Earth that could be detected by another civilization like ours across the Galaxy. We are capable of detecting radio messages from radio telescopes located thousands of light-years away but are no more powerful than ours.

One coded message from Earth was radioed into space in 1974, largely to show off the capabilities of the giant radio telescope at Arecibo, Puerto Rico. This signal was beamed toward the globular cluster M13 in the constellation Hercules, 24,000 light-years away. The least amount of time required to get an answer from M13 at the speed of light is 48,000 years!

Current research concentrates on receiving intelligent radio signals from other civilizations beyond our solar system. This is cheaper, simpler, and safer than deliberately sending out signals until we know where other civilizations are and if they are friendly.

The first attempt to receive intelligent signals was Project Ozma at the U.S. National Radio Astronomy Observatory in Green Bank, West Virginia. Astronomer Frank Drake listened at a frequency of 1420 MHz (21 cm wavelength) to two nearby stars, Tau Ceti and Epsilon Eridani. No intelligent signals were detected then or subsequently.

This lack of success is not at all surprising. Even if other civilizations were deliberately trying to make themselves known to us, we might not have looked in the right direction at the right time, or tuned to the right frequency. The process is rather like trying to find a needle in a haystack by looking only occasionally, and without even knowing what a needle looks like.

About how long would it take a radio-wave message from Earth to reach the Sun's closest neighbor star system, Alpha Centauri (4.3 light-years away)? _____

Answer: 4.3 years. (Radio waves travel at the speed of light.)

12.11 A SERIOUS SEARCH

Scientists can search for extraterrestrial intelligence using existing radio telescopes and computers.

The challenge is to search for a weak, undefined radio signal from an

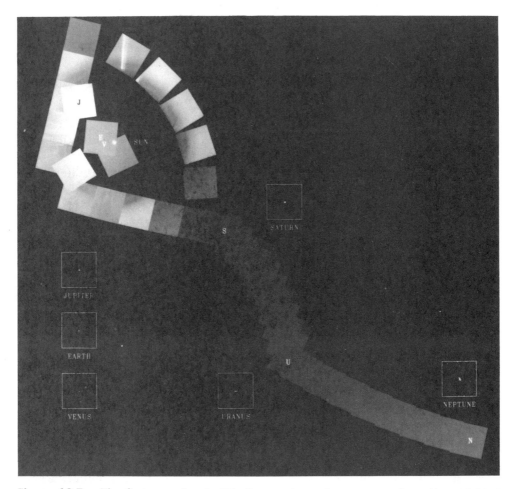

Figure 12.7. The first ever "portrait" of our solar system as seen from the outside. Voyager 1 cameras pointed back and took a series of pictures of the Sun and planets from a distance of approximately 6 billion km (4 billion miles) and 32° above the ecliptic plane on February 14, 1990.

unknown direction, since no one knows exactly how far away or where possible alien transmitters are or which frequencies they use.

The frequencies considered most likely for our first contact are between 1400 MHz and 1700 MHz, often referred to as the galactic "waterhole" where we shall all meet. A modulated signal in this microwave region would stand out because celestial objects emit natural electromagnetic radiation at higher and lower frequencies. Also, a transmitter would require the least power there to generate a detectable signal above the natural background noise.

Hopes rest on affordable automated systems in which computers act as sensitive multichannel spectrum analyzers (MCSA), or radio receivers for a wider band of frequencies that scan millions of radio channels at once.

The most comprehensive proposed search for extraterrestrial civilizations, called the High Resolution Microwave Survey, includes two complementary types of searches:

1. An **all-sky survey** that searches the entire sky over a wide frequency range to detect strong signals. The Deep Space Network's 34-m telescopes in the northern and southern hemispheres scan over frequencies from 1000 MHz to 10,000 MHz and some accessible frequencies to 25,000 MHz.

2. A high-sensitivity **targeted search** that seeks weak signals originating near close stars like our Sun. This targets 1000 stars within 100 light-years of Earth over the frequency range 1000 MHz to 3000 MHz plus any accessible frequencies up to 10,000 MHz.

In the future, if we decide to undertake a more comprehensive search, we can construct even more sensitive multichannel analyzers and bigger antennas or antenna arrays. Alternatives such as putting antennas in space or on the Moon are also possibilities.

Imagine that as a citizen you have been asked to cast a vote for or against our joining a serious international search for signals from other civilizations. How would you vote, and why? _____

_____

_____

_____

_____

Answer: Your answer is a matter of opinion. I would vote for an affordable international search. If we found other intelligent civilizations, they might teach us how to surmount the problems now threatening our survival on Earth. If we did not, the money would still be well spent, since we could expect tremendous gains in peaceful coexistence and knowledge for humanity's benefit from an international commitment to an intellectual and scientific effort of this size.

SELF-TEST

This self-test is designed to show you whether or not you have mastered the material in Chapter 12. Answer each question to the best of your ability. Correct answers and review instructions are given at the end of the test.

1. List two observations that support the theory that the first living things on Earth may have evolved spontaneously out of nonliving chemicals.

   _____

   _____

   _____

   _____

   _____

2. Briefly summarize the scientific theory of the evolution of intelligent life on Earth from simple one-celled organisms. _____

   _____

   _____

   _____

   _____

   _____

   _____

   _____

   _____

   _____

3. Explain why the first search for life on another planet was conducted on Mars. _____

   _____

   _____

   _____

   _____

4. Describe two types of observations that may give evidence for planets circling stars beyond our Sun. (1) _____

_____

_____

(2)_____

_____

_____

_____

_____

5. What is the most uncertain factor involved in estimates of the statistical chances for extraterrestrial intelligent life? _____

_____

6. What is the dominant current scientific view of UFOs?_____

_____

_____

_____

7. Match the following firsts to the correct spacecraft. First:

____ (a) Human in space (Yuri Gagarin).
____ (b) Reusable piloted spacecraft.
____ (c) Spacecraft to leave solar system, bearing a message from Earth.
____ (d) Search for life on surface of another planet.

(1) Space shuttle.
(2) Pioneer 10.
(3) Viking 1 and 2.
(4) Vostok 1.

8. Explain why scientists concentrate on receiving intelligent radio signals in their search for possible life on planets circling other stars beyond our Sun.

_____

_____

_____

_____

_____

9. List two complementary search strategies that are popular today.

(1) _____ ;

_____

_____

(2)_____

_____

_____

ANSWERS

Compare your answers to the questions on the self-test with the answers given below. If you missed any questions, review the sections indicated in parentheses following the answer. If you missed several questions, you should probably reread the entire chapter carefully. Look in the Appendixes for other interesting and useful information about astronomy.

1. (1) Biologists find that all living organisms on Earth depend on a few basic organic molecules. These molecules have been manufactured by energizing gas atoms in laboratories. (2) The common virus has characteristics of both living organisms and nonliving molecules.　(Sections 12.1, 12.3)

2. The principle of natural selection, or the survival of the fittest, asserts that when living species reproduced, variations in characteristics were introduced. Those individuals with favorable variations that helped them to survive the environment were the fittest. They had the best chance of reaching adulthood and reproducing themselves. Thus favorable traits were passed on while unfavorable traits died out, by natural selection. Intelligence is a favorable trait. Over a period of millions of years under the different environmental conditions that existed on Earth, intelligent beings evolved from simple cells.　(Section 12.4)

3. Mars is in the Sun's habitable zone. Evidence suggests that water once flowed on Mars. Certain terrestrial plants and microbes can survive environmental conditions similar to those on Mars. Mars is close enough for the trip to be cost effective.　(Sections 12.5, 12.8, 12.9)

4. (1) Direct observation: A circumstellar disk, which could be a planetary system in its early formative stage. (2) Indirect observation: A perturbation that could be caused by the gravitational tug of massive planets, such as a wobble in the proper motion of a star, a Doppler shift in the spectral lines of a star which indicates radial velocity changes; or small periodic changes in the frequency of radio waves received from a millisecond pulsar.　(Section 12.7)

5. The average lifetime of an intelligent civilization.　(Section 12.6)

6. Most scientists do not think we have been visited by creatures from other worlds. They would be glad to examine concrete evidence, such as a piece of an alien spacecraft, in a laboratory, but none has been found.　(Section 12.9)

7. (a) 4;　(b) 1;　(c) 2;　(d) 3.　(Sections 12.1, 12.8, 12.9)

8. Even the closest stars are several light-years away so we cannot travel there. We have the capability of communicating at the speed of light with other civilizations using radio waves. Concentrating on receiving intelligent signals is cheaper, simpler, and safer than transmitting until we know where other civilizations are and if they are friendly.　(Sections 12.8 through 12.10)

9. (1) All-sky survey that searches the entire sky over a wide frequency range to detect strong signals. (2) High-sensitivity targeted search directed at nearby stars like our Sun, over a narrower frequency range.　(Section 12.11)

EPILOGUE

*Astronomy compels the soul to look upwards
and leads off from this world to another.*

Plato (c. 428–348 B.C.)

The Republic

Astronomy has come a long way since the ancients first pondered the mysteries of the universe. But many exciting discoveries still lie ahead. Now that you have deepened your understanding of basic ideas, you should enjoy observing the sky and contemporary discoveries more than ever!

USEFUL RESOURCES

PERIODICALS

Air and Space, Smithsonian Institution, Subscription Center, P.O. Box 53247, Boulder, CO 80324.

Astronomy, Karmbach Publishing Co., 21027 Crossroads Circle, Waukesha, WI 53187.

The Griffith Observer, Griffith Observatory, 2800 East Observatory Road, Los Angeles, CA 90027.

Mercury, Astronomical Society of the Pacific, 390 Ashton Avenue, San Francisco, CA 94112.

National Geographic, 17th and M Streets, NW, Washington, DC 20036.

Natural History, Membership Services, Box 6000, Des Moines, IA 50340.

Popular Astronomy, 270 Madison Avenue, New York, NY 10016.

Science News, 1719 N Street, NW, Washington, DC 20036.

Scientific American, 415 Madison Avenue, New York, NY 10017.

Sky and Telescope, 49-50-51 Bay State Road, Cambridge, MA 02138.

DIRECTORIES

Sky and Telescope's Astronomy Resource Guide, Annual. Directory of planetariums, observatories, museums, astronomy clubs, societies, dealers, manufacturers, computer bulletin boards, and telephone hotlines in the United States and Canada.

Astronomy's Activity Guide, Annual. Directory of astronomy clubs, events, places, and organizations in the United States and Canada.

American Astronomical Society Membership Directory, Annual. Corporations, publishers, individuals, and North American and international institutions.

BOOKS BY DINAH L. MOCHÉ, Ph.D.

AMAZING ROCKETS, Western Publishing Company, Racine, WI.

AMAZING SPACE FACTS, 2nd edition, Western Publishing Company, Racine, WI

ASTRONOMY, 4th edition, John Wiley & Sons, New York, NY

ASTRONOMY TODAY, updated regularly, Random House, New York, NY

THE ASTRONAUTS, Random House, New York, NY

THE GOLDEN BOOK OF SPACE EXPLORATION, Western Publishing Company, Racine, WI

IF YOU WERE AN ASTRONAUT, 2nd edition, Western Publishing Company Racine, WI

LABORATORY MANUAL FOR INTRODUCTORY ASTRONOMY—editor and co-author, Queensborough Community College Press, Bayside, NY

LIFE IN SPACE, Ridge Press/A & W Publishers, New York, NY

MY FIRST BOOK ABOUT SPACE, Western Publishing Company, Racine, WI

MAGIC SCIENCE TRICKS, Scholastic Book Services, New York, NY

MARS, Franklin Watts, New York, NY

MORE MAGIC SCIENCE TRICKS, Scholastic Book Services, New York, NY

MY FIRST BOOK ABOUT SPACE, Western Publishing Company, Racine, WI

RADIATION, Franklin Watts, New York, NY

SEARCH FOR LIFE BEYOND EARTH, Franklin Watts, New York, NY

WHAT'S UP THERE? QUESTIONS AND ANSWERS ABOUT STARS AND SPACE, updated regularly, Scholastic Book Services, New York, NY

WHAT'S DOWN THERE? QUESTIONS AND ANSWERS ABOUT THE OCEAN, Scholastic Book Services, New York, NY

WE'RE TAKING AN AIRPLANE TRIP, Western Publishing Company, Racine, WI

GENERAL REFERENCE BOOKS

The author has used as principal reference sources for this edition:

Norton's 2000.0, 18th ed., edited by Ian Ridpath (Essex, England: Longman Scientific & Technical, 1989. Co-published in the United States with John Wiley & Sons, Inc., New York). Star atlas, up-to-date astronomical data, and practical observing advice.

Observer's Handbook, edited by Roy L. Bishop, issued annually by The Royal Astronomical Society of Canada, 136 Dupont Street, Toronto, Ontario M5R 1V2. Information and tables on the Sun, Moon, planets, asteroids, meteor showers, and other celestial phenomena.

HANDBOOKS, OBSERVING GUIDES, AND STAR ATLASES

A Field Guide to the Stars and Planets, 2nd edition, by Donald H. Menzel and Jay M. Pasachoff (Boston: Houghton Mifflin Co., 1983).

All About Telescopes by Sam Brown (Barrington, NJ: Edmund Scientific Co.).

The Astronomical Almanac, issued annually by the U.S. Naval Observatory (Washington, DC: U.S. Government Printing Office, yearly). Current information about sun, moon, planets, eclipses, and occultations.

Astrophotography by Robert T. Little (New York: Macmillan Publishing Co., 1986).

Burnham's Celestial Handbook, Volumes 1, 2, and 3 (New York: Dover Publications, Inc., 1978). Observer's guide to space beyond the solar system.

Observational Astronomy Activities: From Naked Eye to Deep Sky by George C. Atamian (2828 E. Foothill Blvd., Pasadena, CA: Bushnell, Division of Bausch & Lomb). Introductory self-instructional guide for observers of all ages.

Sky Atlas 2000.0 by Wil Tirion (Cambridge, MA: Sky Publishing Corp., 1981).

Sky Calendar (East Lansing, MI: Abrams Planetarium, Michigan State University, yearly).

TEACHING MATERIALS

Publication lists, film lists, and information about other services are available from the Educational Services office at the NASA center serving your state. See the list below.

NASA Ames Research Center
Moffett Field, CA 94035
Serves Alaska, Arizona, California, Hawaii, Idaho, Montana, Nevada, Oregon, Utah, Washington, and Wyoming.

NASA Goddard Space Flight Center
Greenbelt, MD 20771
Serves Connecticut, Delaware, District of Columbia, Maine, Maryland, Massachusetts, New Hampshire, New Jersey, New York, Pennsylvania, Rhode Island and Vermont.

NASA Jet Propulsion Laboratory
4800 Oak Grove Drive
Pasadena, CA 91109
Serves inquiries related to space exploration and other JPL activities.

NASA Johnson Space Center
Houston, TX 77058
Serves Colorado, Kansas, Nebraska, New Mexico, North Dakota, Oklahoma, South Dakota, and Texas.

NASA Kennedy Space Center
Kennedy Space Center, FL 32899
Serves Florida, Georgia, Puerto Rico, and the Virgin Islands.

NASA Langley Research Center
Hampton, VA 23665
Serves Kentucky, North Carolina, South Carolina, Virginia, and West Virginia.

NASA Lewis Research Center
Cleveland, OH 44135
Serves Illinois, Indiana, Michigan, Minnesota, Ohio, and Wisconsin.

NASA Marshall Space Flight Center
MSFC, AL 35812
Serves Alabama, Arkansas, Iowa, Louisiana, Missouri, and Tennessee.

NASA Stennis Space Center
Stennis Space Center, MS 39529
Serves Mississippi.

For information about audio visual materials, contact:

NASA CORE
Lorain County Joint Vocation School
15181 Route 58 South
Oberlin, OH 44074

CAREER INFORMATION

A Career in Astronomy, The American Astronomical Society, Education Officer, Mary Kay Hemenway, Ph.D., Dept. of Astronomy, University of Texas, Austin, TX 78712.

Degree Programs in Physics and Astronomy in U.S. Colleges and Universities, American Institute of Physics, 335 East 45th Street, New York, NY 10017.

(a) Physics in Your Future, and *(b) Women in Science* by Dinah L. Moché, Ph.D., American Association of Physics Teachers, 5110 Roanoke Place, Suite 101, College Park, MD 20740. (a) Career information for junior and senior high school, and (b) a multimedia package presenting information about six women scientists and their work. Includes written materials, audio cassette tapes, and 35 mm slides.

ORGANIZATIONS

American Association of Variable Star Observers
25 Birch Street
Cambridge, MA 02138
(617) 354-0484

American Association of Physics Teachers
5112 Berwyn Road
College Park, MD 20740
(301) 345-4200

American Astronomical Society
2000 Florida Avenue, NW, Suite 300
Washington, DC 20009

Astronomical League
6235 Omie Circle
Pensacola, FL 32504
(904) 484-1152

Astronomical Society of the Pacific
390 Ashton Avenue
San Francisco, CA 94112
(415) 337-1100

Astronomy Day Headquarters
Chaffee Planetarium, 54 Jefferson SE
Grand Rapids, MI 49503
(616) 784-9518, (616) 456-3987

British Astronomical Association
Burlington House, Piccadilly
London W1V ONL, England

Earthwatch
Box 403
Watertown, MA 02272
(617) 926-8200

International Planetarium Society
Hansen Planetarium, 15 South State Street
Salt Lake City, UT 84111

National Science Teachers Association
1742 Connecticut Avenue, NW,
Washington, DC 20009
(202) 543-1900

National Space Society
922 Pennsylvania Avenue, SE
Washington, DC 20003
(202) 543-1900

Royal Astronomical Society
Burlington House, Piccadilly
London W1V ONL, England

Royal Astronomical Society of Canada
136 Dupont Street
Toronto, Ontario, Canada, M5R 1V2
(416) 924-7973

Society of Amateur Radio Astronomers
247 North Linden Street
Massapequa, NY 11758
(516) 798-8459

Spaceweek National Headquarters
1110 NASA Road 1, #1100
Houston, TX 77058
(713) 333-3627

The Planetary Society
65 North Catalina Avenue
Pasadena, CA 91106
(818) 793-5100

Western Amateur Astronomers
163 Starlight Crest Drive
La Canada, CA 91011

Appendix 1

THE CONSTELLATIONS

Name and Pronunciation	Genitive	Abbr.	Meaning	Reference RA and Decl.	
Andromeda, ăn-drŏm′ é-dȧ	-dae	And	Daughter of Cassiopeia	1^h	$+40°$
Antlia, ănt lĭ-ȧ	-liae	Ant	The Air Pump	10^h	$-35°$
Apus, ā pŭs	Apodis	Aps	Bird of Paradise	16^h	$-75°$
Aquarius, ȧ-kwâr ĭ-ŭs	-rii	Aqr	The Water-bearer	23^h	$-10°$
Aquila, ăk wĭ-lȧ	-ae	Aql	The Eagle	19^h30^m	$+ 5°$
Ara, ā rȧ	Arae	Ara	The Altar	17^h30^m	$-55°$
Aries, ā rĭ-ēz	Arietis	Ari	The Ram	2^h	$+20°$
Auriga, ô-rī′ gȧ	-gae	Aur	The Charioteer	5^h30^m	$+40°$
Bootes, bô-ō tēz	-tis	Boo	The Herdsman	14^h30^m	$+30°$
Caelum, sē lŭm	Caeli	Cae	The Graving Tool	4^h30^m	$-40°$
Camelopardalis kȧ-měl′ ō-pár′ dȧ-lĭs	Camelopardalis	Cam	The Giraffe	6^h	$+70°$
Cancer, kăn′ sēr	Cancri	Cnc	The Crab	8^h30^m	$+20°$
Canes Venatici kā′ nēz vĕ-năt′ ĭ-sī	Canum Venaticorum	CVn	The Hunting Dogs	12^h30^m	$+40°$
Canis Major, kā′ nĭs māʼ jēr	Canis Majoris	CMa	The Big Dog	7^h	$-20°$
Canis Minor, kā′ nĭs mĭ′ nēr	-ris	CMi	The Little Dog	7^h30^m	$+ 5°$
Capricornus, kăp′ rĭ-kôr′ nŭs	Capricorni	Cap	The Horned Goat	21^h	$-20°$
Carina, kȧ-rī′ nȧ	-nae	Car	The Ship's Keel	9^h	$-60°$
Cassiopeia, kăs′ ĭ-ō-pē′ yȧ	Cassiopeiae	Cas	The Queen	1^h	$+60°$
Centaurus, sĕn-tô′ rŭs	-ri	Cen	The Centaur	13^h	$-50°$
Cepheus, sē′ fŭs	Cephei	Cep	The King	22^h	$+65°$
Cetus, sē′ tŭs	Ceti	Cet	The Whale	2^h	$-10°$
Chamaeleon, kȧ-mē′ lē-ŭn	-ntis	Cha	The Chameleon	10^h	$-80°$
Circinus, sûr′ sĭ-nŭs	-ni	Cir	The Compasses	15^h	$-60°$
Columba, kō-lŭm′ bȧ	-bae	Col	The Dove	6^h	$-35°$
Coma Berenices kō′ mȧ běr′ ē-nĭ sēz	Comae Berenices	Com	Berenice's Hair	13^h	$+25°$
Corona Australis kō-rō′ nȧ ôs-trā′ lĭs	-nae	CrA	The Southern Crown	19^h	$-40°$
Corona Borealis kō-rō′ nȧ bō′ rē-ā′ lĭs	Coronae Borealis	CrB	The Northern Crown	15^h30^m	$+30°$
Corvus, kôr′ vŭs	Corvi	Crv	The Crow	12^h30^m	$-20°$
Crater, krä′ tēr	-ris	Crt	The Cup	11^h30^m	$-15°$
Crux, krŭks	Crucis	Cru	The Southern Cross	12^h30^m	$-60°$
Cygnus, sĭg′ nŭs	Cygni	Cyg	The Swan	20^h	$+40°$
Delphinus, dĕl-fĭ′ nŭs	-ni	Del	The Dolphin	20^h30^m	$+15°$
Dorado, dō-rä′ dō	-dus	Dor	The Swordfish	5^h	$-60°$
Draco, drä′ kō	Draconis	Dra	The Dragon	18^h	$+70°$
Equuleus, ē-kwōō′ lē-ŭs	-lei	Equ	The Colt	21^h	$+10°$
Eridanus, ē-rĭd′ ȧ-nŭs	Eridani	Eri	The River	3^h30^m	$-20°$
Fornax, tôr′ năks	-nacis	For	The Furnace	3^h	$-30°$
Gemini, jěm′ ĭ-nī	Germinorum	Gem	The Twins	7^h	$+25°$
Grus, grŭs	Gruis	Gru	The Crane	22^h	$-40°$
Hercules, hûr′ kū-lēz	Herculis	Her	Hercules	17^h	$+35°$
Horologium, hŏr′ ō-lō′ jĭ-ŭm	-gii	Hor	The Clock	3^h	$-50°$
Hydra, hī′ drȧ	Hydrae	Hya	The Water Snake (female)	11^h	$-20°$
Hydrus, hī′ drŭs	-dri	Hyi	The Water Snake (male)	2^h	$-70°$
Indus, ĭn′ dŭs	Indi	Ind	The Indian	21^h	$-50°$
Lacerta, la-sûr′ tȧ	-tae	Lac	The Lizard	22^h30^m	$+50°$
Leo, lē′ ō	Leonis	Leo	The Lion	10^h30^m	$+20°$
Leo Minor, lē′ ō mĭ′ nēr	Leonis Minoris	LMi	The Little Lion	10^h30^m	$+35°$
Lepus, lē′ pŭs	Leporis	Lep	The Hare	5^h30^m	$-20°$
Libra, lī′ bra	Librae	Lib	The Balance	15^h	$-15°$
Lupus, lū′ pŭs	Lupi	Lup	The Wolf	15^h30^m	$-40°$
Lynx, lĭnks	Lyncis	Lyn	The Lynx	8^h	$+50°$
Lyra, lī′ rȧ	Lyrae	Lyr	The Lyre	18^h30^m	$+35°$
Mensa, měn′ sȧ	-sae	Men	Table Mountain	5^h30^m	$-75°$
Microscopium mĭ′ krō-skō′ pĭ-ŭm	-pii	Mic	The Microscope	21^h	$-35°$

(continued)

Name and Pronunciation	Genitive	Abbr.	Meaning	Reference RA and Decl.	
Monoceros, mō-nŏs′ ēr-ŏs	Monocerotis	Mon	The Unicorn	7^h	$- 5°$
Musca, mŭs′ kȧ	-cae	Mus	The Fly	12^h	$-70°$
Norma, nôr′ ma	-mae	Nor	The Square	16^h	$-50°$
Octans, ŏk′ tănz	-ntis	Oct	The Octant	0^h-25^h	$-90°$
Ophiuchus, ŏf′ ĭ-ū′ kŭs	Ophiuchi	Oph	The Serpent-bearer	17^h	$0°$
Orion, ō-rī′ ŏn	Orionis	Ori	The Hunter	5^h30^m	$0°$
Pavo, pā′ vō	-vonis	Pav	The Peacock	20^h	$-65°$
Pegasus, pĕg′ ȧ-sŭs	Pegasi	Peg	The Winged Horse	23^h30^m	$+20°$
Perseus, pûr′ sūs	Persei	Per	Perseus	3^h30^m	$+45°$
Phoenix, fē′ nĭks	Phoenicis	Phe	The Phoenix	1^h	$-50°$
Pictor, pĭk′ tēr	Pictoris	Pic	The Painter's Easel	6^h	$-55°$
Pisces, pĭs′ ēz	Piscium	Psc	The Fishes	23^h30^m	$+ 5°$
Piscis Austrinus pĭs′ is ôs-trī nŭs	Piscis Austrini	PsA	The Southern Fish	23^h	$-30°$
Puppis, pŭp′ ĭs	Puppis	Pup	The Ship's Stern	8^h	$-40°$
Pyxis, pĭk′ sĭs	Pyxidis	Pyx	The Compass	9^h	$-30°$
Reticulum, rē-tĭk′ ū-lŭm	-li	Ret	The Net	4^h	$-60°$
Sagitta, sȧ-jĭt′ ȧ	-tae	Sge	The Arrow	20^h	$+20°$
Sagittarius, săj′ ĭ-tā′ rĭ-ŭs	Sagittarii	Sgr	The Archer	18^h30^m	$-30°$
Scorpius, skôr′ pĭ-ŭs	Scorpii	Sco	The Scorpion	17^h	$-30°$
Sculptor, skŭlp′ tēr	-ris	Scl	The Sculptor	0^h	$-30°$
Scutum, skū′ tŭm	Scuti	Sct	The Shield	18^h30^m	$-10°$
Serpens, sûr′ pĕnz	Serpentis	Ser	The Serpent	16^h	$0°$
Sextans, sĕks′ tănz	-ntis	Sex	The Sextant	10^h	$- 5°$
Taurus, tô′ rŭs	Tauri	Tau	The Bull	4^h30^m	$+15°$
Telescopium, tĕl′ ē-skō′ pĭ-ŭm	-pii	Tel	The Telescope	19^h	$-50°$
Triangulum, trī-ăng′ gū-lŭm	Trianguli	Tri	The Triangle	2^h	$+30°$
Triangulum Australe trī-ăng′ gū-lŭm ôs-trä′ lē	Trianguli Australis	TrA	The Southern Triangle	16^h	$-65°$
Tucana, tū-kā′ nȧ	-nae	Tuc	The Toucan	23^h30^m	$-65°$
Ursa Major, ûr′ sȧ mā′ jēr	Ursae Majoris	UMa	The Great Bear	11^h	$+60°$
Ursa Minor, ûr′ sȧ mī′ nēr	Ursae Minoris	UMi	The Little Bear	15^h	$+70°$
Vela, vē′ lȧ	Velorum	Vel	The Ship's Sails	9^h	$-50°$
Virgo, vûr′ gō	Virginis	Vir	The Maiden	13^h	$-10°$
Volans, vō′ lănz	-ntis	Vol	The Flying Fish	8^h	$-70°$
Vulpecula, vŭl-pĕk′ ū-lȧ	Vulpeculae	Vul	The Fox	19^h30^m	$+25°$

ā dāte; ă tăp; â câre; ȧ ȧsk; ē wē; ĕ mĕt; ē makēr; ī īce; ĭ bĭt; ō gō; ŏ hŏt; ô ôrb; o͞o mo͞on; ū, ūnite; ŭ ŭp; û ûrn.

Velocity of light	c	= 299,792,458 meters per second
Gravitation constant	G	= 6.672×10^{-11} Nm^2 kg^{-2}
Stefan-Bolzmann constant	σ	= 5.67×10^{-8} Wm^{-2} K^{-4}
Mass of electron	m_e	= 9.1094×10^{-31} kg
Mass of hydrogen atom	m_H	= 1.67352×10^{-24} gram
Mass of proton	m_p	= 1.67262×10^{-27} kg
Astronomical unit	AU	= $1.49597870 \times 10^{11}$ m
Parsec	pc	= 3.085678×10^{16} m
		= 3.261631 light-years
Light-year	LY	= 9.460536×10^{15} m
Mass of sun	$M_\odot$	= 1.9891×10^{30} kg
Radius of sun	**$R_\odot$	= 696,265 km
Solar radiation	$L_\odot$	= 3.85×10^{26} W
Mass of Earth	***$M_\oplus$	= 5.974×10^{24} kg
Equatorial radius of Earth	$R_\oplus$	= 6,378.140 km
Direction of Galactic center	RA	= 264°.83, DEC = 28°.90 (1900)
Ephemeris day	d_E	= 86,400 seconds
Tropical year (equinox to equinox)		= 365.2422 ephemeris days
Sidereal year		= 365.2564 ephemeris days

** $\odot$ = of the Sun
*** $\oplus$ = of the Earth

Source: Data reprinted by permission of *The Royal Astronomical Society of Canada Observer's Handbook (1993).*

MEASUREMENTS AND SYMBOLS

METRIC AND U.S. MEASURES

Approximate Conversions to U.S. Units

Symbol	When You Know	Multiply By	To Find	Symbol
		Length		
cm	centimeters	0.39	inches	in
cm	centimeters	0.03	feet	ft
m	meters	1.1	yards	yd
km	kilometers	0.6	miles	ml
		Area		
cm^2	squarc centimeter	0.16	square inches	in^2
m^2	square meters	11	square feet	ft^2
m^2	square meters	1.2	square yards	yd^2
km^2	square kilometer	0.4	square miles	ml^2
		Mass (weight)		
g	grams	0.03	ounces	oz
kg	kilograms	2.2	pounds	lb
t	metric ton	1.1	short tons (2000 lb)	
		Volume		
L	liters	0.26	gallons	gal
m^3	cubic meters	35	cubic feet	ft^3
m^3	cubic meters	1.3	cubic yards	yd^3
		Temperature (exact)		
°C	degrees Celsius	9/5°C + 32	degrees Fahrenheit	°F
K	degrees Kelvin	9/5 K – 460	degrees Fahrenheit	°F

ANGULAR MEASURE

A circle contains 360 degrees, or 360°.
1° contains 60 minutes of arc, or 60'.
1' contains 60 seconds of arc, or 60".

THE GREEK ALPHABET

Alpha	A	α	Nu	N	ν
Beta	B	β	Xi	Ξ	ξ
Gamma	Γ	γ	Omicron	O	o
Delta	Δ	δ	Pi	Π	π
Epsilon	E	ε	Rho	P	ρ
Zeta	Z	ζ	Sigma	Σ	σ
Eta	H	η	Tau	T	τ
Theta	Θ	θ, φ	Upsilon	Υ	υ
Iota	I	ι	Phi	Φ	φ, φ
Kappa	K	κ	Chi	X	χ
Lambda	Λ	λ	Psi	Ψ	ψ
Mu	M	μ	Omega	Ω	ω

SUN, MOON, PLANET SYMBOLS

☉	Sun	♂	Mars	P	Pluto
☾	Moon	♃	Jupiter	●	new moon
☿	Mercury	♄	Saturn	☽	first quarter
♀	Venus	⛢	Uranus	○	full moon
⊕	Earth	♆	Neptune	☾	last quarter

SIGNS OF THE ZODIAC

♈	Aries	0°	♌	Leo	120°	♐	Sagittarius	240°
♉	Taurus	30°	♍	Virgo	150°	♑	Capricornus	270°
♊	Gemini	60°	♎	Libra	180°	♒	Aquarius	300°
♋	Cancer	90°	♏	Scorpius	210°	♓	Pisces	330°

Appendix 4

PERIODIC TABLE OF THE ELEMENTS

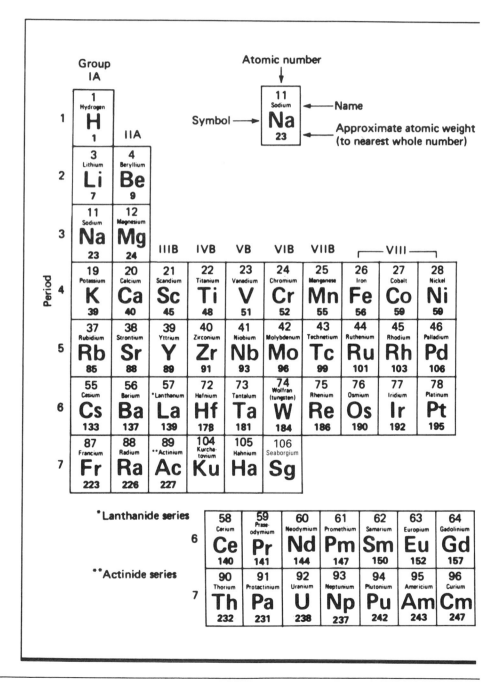

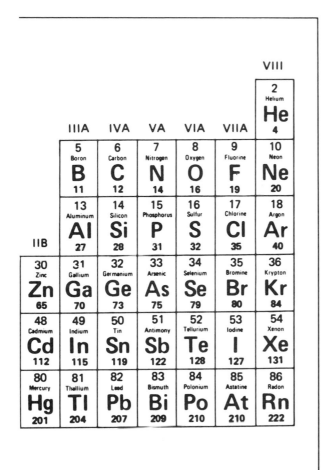

Appendix 5

THE NEAREST STARS

Star	RA 2000.0 h m	Dec. ° '	Apparent Magnitude	Spectral Class	Parallax "	Distance (light-years)	Absolute Magnitude
Sun	—	—	−26.72	G2	—	—	4.8
Proxima (V645 Cen)	14 29.7	−62 41	11.05 (var.)	M5.5	0.772	4.2	15.5
α Cen A	14 39.6	−60 50	−0.01	G2	0.750	4.3	4.4
B			1.33	K1			5.7
Barnard's Star	17 57.8	+04 34	9.54	M3.8	0.545	6.0	13.2
Wolf 359 (CN Leo)	10 56.5	+07 01	13.53 (var.)	M5.8	0.421	7.7	16.7
BD +36°2147	11 03.3	+35 58	7.50	M2.1	0.397	8.2	10.5
UV Cet A	01 38.8	−17 57	12.52 (var.)	M5.6	0.387	8.4	15.5
B			13.02 (var.)	M5.6			16.0
Sirius A	06 45.1	−16 43	−1.46	A1	0.377	8.6	1.4
B			8.3	DA			11.2
Ross 154	18 49.8	−23 50	10.45	M3.6	0.345	9.4	13.1
Ross 248	23 41.9	+44 10	12.29	M4.9	0.314	10.4	14. 8
ε Eri	03 32.9	−09 28	3.73	K2	0.303	10.8	6.1
Ross 128	11 47.8	+00 48	11.10	M4.1	0.298	10.9	13. 5
61 Cyg A (V1803 Cyg)	21 06.9	+38 45	5.22 (var.)	K3.5	0.294	11.1	7.6
B			6.03	K4. 7			8.4
ε Ind	22 03.4	−56 47	4.68	K3	0.291	11.2	7.0
BD +43°44 A	00 18.5	+44 01	8.08	M1.3	0.290	11.2	10.4
B			11.06	M3.8			13.4
L789-6	22 38.5	−15 19	12.18		0.290	11.2	14.5
Procyon A	07 39.3	+05 13	0.38	F51	0.285	11.4	2.6
B			10.7	DF			13.0
BD +59°1915 A	18 43.1	+59 38	8.90	M3.0	0.282	11.6	11.2
B			9.69	M3.5			11.9
CoD −36°15693	23 05.9	−35 51	7.35	M1.3	0.279	11.7	9.6

Source: Alan Batten, *Royal Astronomical Society of Canada Observer's Handbook* (1989).

Appendix 6

THE MESSIER OBJECTS

Messier No. (M)	NGC	RA		Dec		Constellation	Description
		h	m	°	'		
1	1952	5	34.5	+22	01	Tau	Supernova remnant
2	7089	21	33.5	−0	49	Aqr	Globular cluster
3	5272	13	42.2	+28	23	CVn	Globular cluster
4	6121	16	23.6	−26	32	Sco	Globular cluster
5	5904	15	18.6	+2	05	Ser	Globular cluster
6	6405	17	40.1	−32	13	Sco	Open cluster
7	6475	17	53.9	−34	49	Sco	Open cluster
8	6523	18	03.8	−24	23	Sgr	Diffuse nebula
9	6333	17	19.2	−18	31	Oph	Globular cluster
10	6254	16	57.1	−4	06	Oph	Globular cluster
11	6705	18	51.1	−6	16	Sct	Open cluster
12	6218	16	47.2	−1	57	Oph	Globular cluster
13	6205	16	41.7	+36	28	Her	Globular cluster
14	6402	17	37.6	−3	15	Oph	Globular cluster
15	7078	21	30.0	+12	10	Peg	Globular cluster
16	6611	18	18.8	−13	47	Ser	Open cluster
17	6618	18	20.8	−16	11	Sgr	Diffuse nebula
18	6613	18	19.9	−17	08	Sgr	Open cluster
19	6273	17	02.6	−26	16	Oph	Globular cluster
20	6514	18	02.6	−23	02	Sgr	Diffuse nebula
21	6531	18	04.6	−22	30	Sgr	Open cluster
22	6656	18	36.4	−23	54	Sgr	Globular cluster
23	6494	17	56.8	−19	01	Sgr	Open cluster
24		18	16.9	−18	29	Sgr	*See notes*
25	IC4725	18	31.6	−19	15	Sgr	Open cluster
26	6694	18	45.2	−9	24	Sct	Open cluster
27	6853	19	59.6	+22	43	Vul	Planetary nebula
28	6626	18	24.5	−24	52	Sgr	Globular cluster
29	6913	20	23.9	+38	32	Cyg	Open cluster
30	7099	21	40.4	−23	11	Cap	Globular cluster
31	224	0	42.7	+41	16	And	Spiral galaxy
32	221	0	42.7	+40	52	And	Elliptical galaxy
33	598	1	33.9	+30	39	Tri	Spiral galaxy
34	1039	2	42.0	+42	47	Per	Open cluster
35	2168	6	08.9	+24	20	Gem	Open cluster
36	1960	5	36.1	+34	08	Aur	Open cluster
37	2099	5	52.4	+32	33	Aur	Open cluster
38	1912	5	28.7	+35	50	Aur	Open cluster

(continued)

Messier No. (M)	NGC	RA		Dec		Constellation	Description
		h	m	°	′		
39	7092	21	32.2	+48	26	Cyg	Open cluster
40		12	22.4	+58	05	UMa	*See notes*
41	2287	6	47.0	−20	44	CMa	Open cluster
42	1976	5	35.4	−5	27	Ori	Diffuse nebula
43	1982	5	35.6	−5	16	Ori	Diffuse nebula
44	2632	8	40.1	+19	59	Cnc	Open cluster
45		3	47.0	+24	07	Tau	Open cluster
46	2437	7	41.8	−14	49	Pup	Open cluster
47	2422	7	36.6	−14	30	Pup	Open cluster
48	2548	8	13.8	−5	48	Hya	Open cluster
49	4472	12	29.8	+8	00	Vir	Elliptical galaxy
50	2323	7	03.2	−8	20	Mon	Open cluster
51	5194-5	13	29.9	+47	12	CVn	Spiral galaxy
52	7654	23	24.2	+61	35	Cas	Open cluster
53	5024	13	12.9	+18	10	Com	Globular cluster
54	6715	18	55.1	−30	29	Sgr	Globular cluster
55	6809	19	40.0	−30	58	Sgr	Globular cluster
56	6779	19	16.6	+30	11	Lyr	Globular cluster
57	6720	18	53.6	+33	02	Lyr	Planetary nebula
58	4579	12	37.7	+11	49	Vir	Spiral galaxy
59	4621	12	42.0	+11	39	Vir	Elliptical galaxy
60	4649	12	43.7	+11	33	Vir	Elliptical galaxy
61	4303	12	21.9	+4	28	Vir	Spiral galaxy
62	6266	17	01.2	−30	07	Oph	Globular cluster
63	5055	13	15.8	+42	02	CVn	Spiral galaxy
64	4826	12	56.7	+21	41	Com	Spiral galaxy
65	3623	11	18.9	+13	05	Leo	Spiral galaxy
66	3627	11	20.2	+12	59	Leo	Spiral galaxy
67	2682	8	50.4	+11	49	Cnc	Open cluster
68	4590	12	39.5	−26	45	Hya	Globular cluster
69	6637	18	31.4	−32	21	Sgr	Globular cluster
70	6681	18	43.2	−32	18	Sgr	Globular cluster
71	6838	19	53.8	+18	47	Sge	Globular cluster
72	6981	20	53.5	−12	32	Aqr	Globular cluster
73	6994	20	58.9	−12	38	Aqr	*See notes*
74	628	1	36.7	+15	47	Psc	Spiral galaxy
75	6864	20	06.1	−21	55	Sgr	Globular cluster
76	650-1	1	42.4	+51	34	Per	Planetary nebula
77	1068	2	42.7	−0	01	Cet	Spiral galaxy
78	2068	5	46.7	+0	03	Ori	Diffuse nebula
79	1904	5	24.5	−24	33	Lep	Globular cluster
80	6093	16	17.0	−22	59	Sco	Globular cluster
81	3031	9	55.6	+69	04	UMa	Spiral galaxy
82	3034	9	55.8	+69	41	UMa	Irregular galaxy
83	5236	13	37.0	−29	52	Hya	Spiral galaxy
84	4374	12	25.1	+12	53	Vir	Elliptical galaxy
85	4382	12	25.4	+18	11	Com	Elliptical galaxy
86	4406	12	26.2	+12	57	Vir	Elliptical galaxy

Messier No. (M)	NGC	RA		Dec		Constellation	Description
		h	m	°	′		
87	4486	12	30.8	+12	24	Vir	Elliptical galaxy
88	4501	12	32.0	+14	25	Com	Spiral galaxy
89	4552	12	35.7	+12	33	Vir	Elliptical galaxy
90	4569	12	36.8	+13	10	Vir	Spiral galaxy
91	4548	12	35.4	+14	30	Com	Spiral galaxy
92	6341	17	17.1	+43	08	Her	Globular cluster
93	2447	7	44.6	−23	52	Pup	Open cluster
94	4736	12	50.9	+41	07	CVn	Spiral galaxy
95	3351	10	44.0	+11	42	Leo	Spiral galaxy
96	3368	10	46.8	+11	49	Leo	Spiral galaxy
97	3587	11	14.8	+55	01	UMa	Planetary nebula
98	4192	12	13.8	+14	54	Com	Spiral galaxy
99	4254	12	18.8	+14	25	Com	Spiral galaxy
100	4321	12	22.9	+15	49	Com	Spiral galaxy
101	5457	14	03.2	+54	21	UMa	Spiral galaxy
102							*See notes*
103	581	1	33.2	+60	42	Cas	Open cluster
104	4594	12	40.0	−11	37	Vir	Spiral galaxy
105	3379	10	47.8	+12	35	Leo	Elliptical galaxy
106	4258	12	19.0	+47	18	CVn	Spiral galaxy
107	6171	16	32.5	−13	03	Oph	Globular cluster
108	3556	11	11.5	+55	40	UMa	Spiral galaxy
109	3992	11	57.6	+53	23	UMa	Spiral galaxy
110	205	0	40.4	+41	41	And	Elliptical galaxy

M1	Crab Nebula
M8	Lagoon Nebula; contains a star cluster
M11	Wild Duck Cluster
M16	Surrounded by the Eagle Nebula
M17	Omega Nebula
M20	Trifid Nebula
M24	Star field in Sagittarius, containing the open cluster NGC 6603
M27	Dumbbell Nebula
M31	Andromeda Galaxy
M40	Faint double star Winnecke 4, mags. 9.0 and 9.6
M42, M43	Orion Nebula
M44	Praesepe, the Beehive Cluster
M45	The Pleiades; no NGC or IC number
M51	Whirlpool Galaxy
M57	Ring Nebula
M64	Black Eye Galaxy
M73	Small group of four faint stars
M97	Owl Nebula
M102	Duplicate of M101
M104	Sombrero Galaxy

Source: A. Hirshfeld and R. W. Sinnott (eds.), *Sky Catalogue 2000.0*, Vol. 2 (Sky Publishing Corp./Cambridge University Press, 1985).

INDEX

Note: Bold entries indicate definition of term.

SPRING SKIES

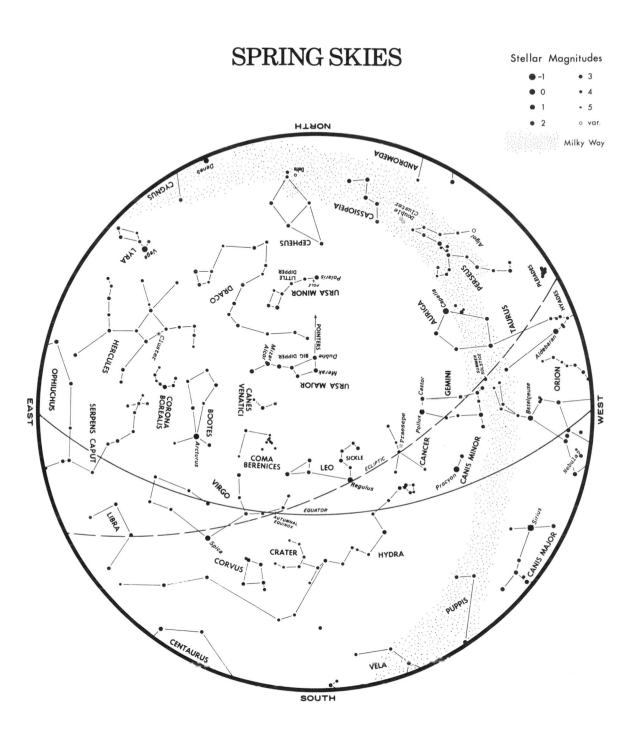

Chart by George Lovi

TIME SCHEDULE

Late March	11 p.m.	Early May	8 p.m.
Early April	10 p.m.	Late May	7 p.m.
Late April	9 p.m.	Early June	6 p.m.

STANDARD TIME

SUMMER SKIES

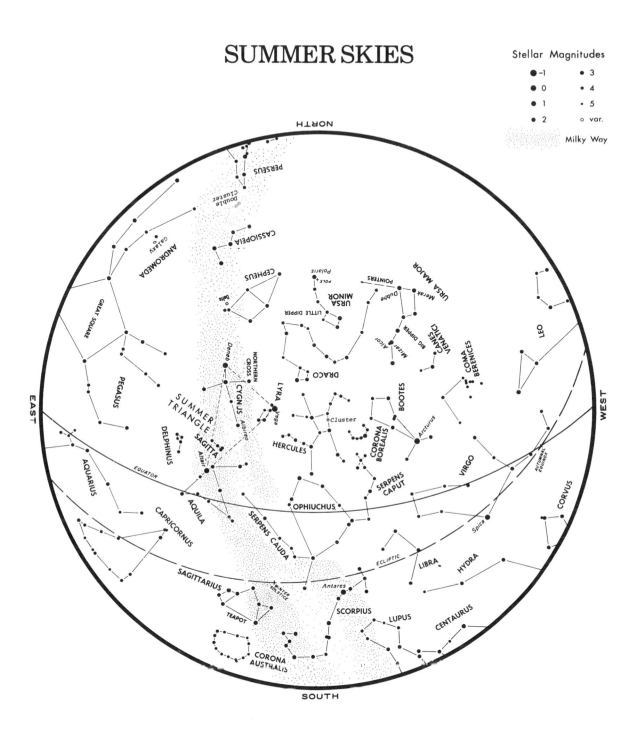

Stellar Magnitudes

- ● −1
- ● 0
- ● 1
- ● 2
- ● 3
- ● 4
- · 5
- ○ var.

Milky Way

Chart by George Lovi

TIME SCHEDULE

Late June	11 p.m.	Early August	8 p.m.
Early July	10 p.m.	Late August	7 p.m.
Late July	9 p.m.	Early September	6 p.m.

STANDARD TIME

AUTUMN SKIES

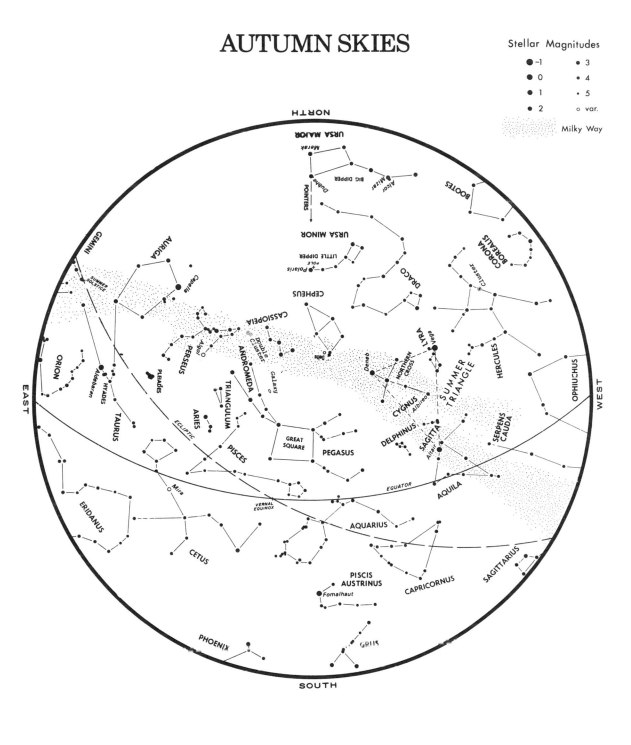

Chart by George Lovi

TIME SCHEDULE

Late September	11 p.m.	Early November	8 p.m.
Early October	10 p.m.	Late November	7 p.m.
Late October	9 p.m.	Early December	6 p.m.

STANDARD TIME

WINTER SKIES

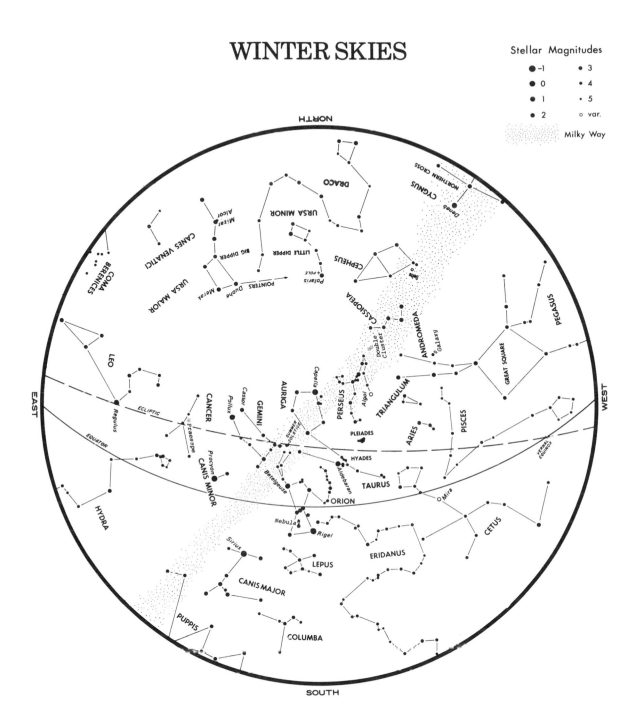

TIME SCHEDULE

Chart by George Lovi

Late December	11 p.m.	Early February	8 p.m.
Early January	10 p.m.	Late February	7 p.m.
Late January	9 p.m.	Early March	6 p.m.

STANDARD TIME

MOON MAP

Chart by George Lovi